Environmental Exposures and Population Health: How Epidemiology Can Guide Public Health Action

Chekhov

Table of Contents

Chapter 1: Introduction

1.1 Environmental Epidemiology

Environmental epidemiology is a branch of epidemiology and public health research that investigates how environmental hazards, whether naturally occurring (e.g., radon, metals), or human-created (e.g., industrial chemicals, air pollution), impact human health. The US National Research Council defines environmental epidemiology as "the study of the effect on human health of physical, biological, and chemical factors in the external environment (National Research Council, 1991) .

Environmental epidemiology focuses on exposure assessment and measurement, and how exposure is associated with population-wide health impacts (Rothman, 1993) Environmental epidemiology is similar to occupational epidemiology in that unlike clinical or infectious disease epidemiology, exposures in environmental epidemiology are often low dose and long-term or chronic, which can pose significant challenges for measurement and assessment (P. O. Wilkinson, 2006). Exposure assessment can be performed using large-scale estimates based on the best available data (e.g., air pollution monitoring used as an estimate of biological markers for individual exposures to certain pollutants), or by using direct measurements of biomarkers from participants' bodies (Needham et al., 2007). The number of environmental epidemiological cohort studies has grown over the last 30 years, based largely on the improved technological capacity to directly measure biomarkers of chemical exposure. Ensuring appropriate classification of exposure remains one of the major challenges in environmental epidemiology, but as laboratory equipment and novel techniques of measurement continue to improve, the quality of data measured improves, and the participant burden can be reduced.

Challenges in environmental epidemiology are similar to those in other types of epidemiology. Most environmental epidemiology studies are non-experimental, observational research, including a large number of cross-sectional studies, case control studies, and, more recently, large cohort studies based on routinely collected data (Rothman, 1993). While longitudinal cohort studies in particular can provide good evidence for quantifying environmental risk, high quality cohort studies are expensive and time-intensive; it is also challenging to recruit and retain participants. In addition, data from these studies may be plagued by unmeasured confounding and measurement error.

While environmental epidemiology has many challenges, it has also contributed to several significant public health successes. The discovery that there is no safe level of lead exposure for children is a major public health finding with population-level impacts (Health Canada, 2013; Lanphear et al., 2005). The discovery of adverse health effects associated with substances such as Bisphenol A (Health Canada, 2008; European Chemicals Agency, 2017) and mercury (Grandjean et al., 1997) also point to the importance of this kind of work. Our study on phthalates aims to build on established best practices for environmental epidemiological investigations and incorporate a relatively novel statistical technique to solve modern challenges related to chemical mixture exposure.

1.2 Problem Statement and Objectives

Phthalates are a chemical group of human-manufactured aromatic esters that are found in a wide variety of consumer goods (Health Canada, 2017). Phthalate exposure is ubiquitous in Canada (Health Canada, 2010). At high doses, phthalates act as endocrine disruptors and are associated with a variety of endocrine sensitive outcomes (Diamanti-Kandarakis et al., 2009). In humans, some evidence indicates that phthalate exposure during endocrine-sensitive periods, including pregnancy and early childhood, may be associated with adverse outcomes in hormone-sensitive processes, including a variety of neurodevelopmental outcomes including Attention Deficit/Hyperactivity Disorder (ADHD), Intelligence Quotient (IQ) scores, a number of autistic traits, and measures of fine motor control (Daniel et al., 2020; Doherty et al., 2017; Engel et al., 2018; Factor-Litvak et al., 2014; Olesen et al., 2018; Oulhote et al., 2020; Polanska et al., 2014; Téllez-Rojo et al., 2013; Testa et al., 2012). Phthalates can cross the placenta and alter endocrine sensitive outcomes in the developing fetus (Mitro, Johnson, & Zota, 2015). The relationship between fetal exposure to phthalates and childhood IQ scores has been investigated in several birth cohort studies (Factor-Litvak et al., 2014; H. B. Huang et al., 2015; Hyland et al., 2019; Jankowska et al., 2019; J. I. Kim et al., 2017; Nakiwala et al., 2018; E. M. Tanner et al., 2019) but these studies did not consider the potential effects of exposure to phthalate mixtures. From a biological perspective, phthalates are believed to be endocrine disruptors, and exposure to groups of phthalates may result in adverse health effects, even if single-chemical exposures are low (Kortenkamp, 2014)

In this study, I investigated the association between prenatal phthalate exposure during pregnancy and Child IQ scores at age 3 in the Maternal-Infant Research on Environmental Chemicals cohort (MIREC) and the follow up MIREC Child Development (MIREC-CD Plus) study. The objectives of this study are to investigate the association between individual phthalate metabolites measured during the first trimester of gestation, and the child's IQ measured at age 3. A secondary objective of this study is to investigate potential sex differences in the association between phthalate exposure and IQ scores. The final objective of my study is to investigate whether exposure to a mixture of phthalates is associated with child IQ, using a method called weighted quantile sum regression (WQSR). This study aims to consider potential confounding using multivariable linear regression, and employs model-building techniques to reduce bias.

Chapter 1 of this book will provide an introduction and rationale for the study. Chapter 2 consists of a literature review to properly situate the problem in the current academic environment. Chapter 3 describes the cohort study that was the data source for this project; Chapter 4 is the prepared manuscript, ready for submission to academic journals. Chapter 5 provides further discussion of the results that are provided in Chapter 4, as well as suggestions for further directions for this type of study. This project represents a secondary analysis of data, with high quality and novel statistical techniques.

Phthalates are a group of chemicals that are regularly used in industrial and consumer products (Heudorf et al., 2007; Schettler et al., 2006). Initially, phthalates were considered safe due to their rapid breakdown in the human body, but further testing on animal models showed that there are adverse effects associated with repeated exposures to the chemicals (Gray & Gangolli, 1986; Lloyd & Foster, 1988). With improvements in biomonitoring science, epidemiological studies monitoring human exposure to phthalates became more feasible and common.

Several studies have reported an inverse association between phthalate exposure and neurodevelopmental outcomes in young children, including language development (Olesen et al., 2018) IQ scores (Factor-Litvak et al., 2014), and psychomotor and behavioural development (Engel et al., 2018; Testa et al., 2012; Whyatt et al., 2012) The inverse relationship between phthalates and neurodevelopmental outcomes is hypothesized to be due to endocrine disruption during the early gestational period. Phthalates may interfere with thyroid hormones, essential for normal brain development during gestation (X. Dong et al., 2017; Ghisari & Bonefeld-Jorgensen, 2009; Poon et al., 1997; Sugiyama et al., 2005). Alternately, phthalates have also been found to act as antiandrogens (Andrade et al., 2006; Christen et al., 2012; Howdeshell et al., 2008) and higher exposure may interfere with sex-specific organization of neural structures during early life development. Hormone disruption during the first trimester of pregnancy may be particularly important to consider because the first trimester is a critical period of growth and reorganization for many neural structures (Buss et al., 2012; Kostović et al., 2002; Vohr et al., 2017) and a time period during which a developing fetus relies exclusively on maternal sources of thyroid hormones (Shepard, 1967). A wide variety of studies have examined the association between prenatal phthalate exposure and IQ scores (Factor-Litvak et al., 2014; H. B. Huang et al., 2015; Hyland et al., 2019; Jankowska et al., 2019; J. I. Kim et al., 2017; Nakiwala et al., 2018; E Tanner et al., 2019). Traditionally, studies that look at phthalate exposure and IQ scores do so using a single exposure-single outcome approach, and traditional regression modelling. More recently, environmental epidemiologists have begun to investigate the effects of chemical mixtures on humans (Joseph M Braun et al., 2016; Carlin et al., 2013). Toxicologists have found some evidence from animal studies that phthalates act in a biologically additive way in the body (Howdeshell et al., 2008)which may mean that low-level exposure to individual chemicals may be harmful when the chemicals are considered together.

My book aims to add to existing literature on phthalate exposure and intellectual ability using a pan-Canadian cohort, which will be the first investigation of this type in a Canadian population. My book also aims to help fill the data gap in assessing correlated phthalates as a mixture, using a novel statistical technique that is still relatively new in environmental epidemiology.

1.4 Public Health Significance

Phthalates are chemicals that most Canadians come in contact with on a daily basis (Health Canada, 2010). Since 2007, the same 11 phthalates metabolites that were measured in MIREC (MnBP, MEP, MBzP, MCHP, MEHP, MOP, MiNP, MMP, MCPP, MEHHP, and MEOHP) have been measured as a part of the ongoing Canadian Health Measures Survey (CHMS), a national biomonitoring and physical examination survey sampled to provide representative data (Cycle 1, 2007-2009, Cycle 2 2009-2011, and Cycle 5, 2016-2017). In the CHMS, all participants had detectable levels of at least one phthalate metabolite in urine samples (Haines et al., 2017; Health Canada, 2019). For this reason, adverse health effects associated with phthalates are potentially relevant to all Canadians. My book focuses specifically on early life and development, and will provide additional data for assessment of how phthalates may affect intelligence scores during an important window of development. These results may be included in future risk assessments, particularly when considering cumulative phthalate exposure.

From a public health perspective, higher childhood IQ scores have consistently been linked with reduced all-cause mortality in adulthood in large cohorts in Scotland (Calvin et al., 2017; Warrilow et al., 2021), Sweden (Sörberg Wallin et al., 2018), Britian, United States (Martin & Kubzansky, 2005) and in at least two meta-analyses (Calvin et al., 2011; Dobson et al., 2017). Higher IQ scores has also been linked to a reduced number of chronic health conditions in adulthood (Wraw et al., 2015), even when factors such as socio-economic status and childhood adversity are considered. Additionally, studies estimate that the societal cost of deficits in intellectual abilities can amount to billions of dollars (Attina et al., 2016; Gaylord et al., 2020; Trasande et al., 2015).

Chapter 2: Literature Review

2.1 Phthalate Chemistry and Environmental Sources

2.1.1 Chemical and Physical Properties of Phthalates

The chemical group known as phthalates is composed of diesters of 1, 2-benzenedicarboxylic acid (NRC, 2008). Chemically, phthalates are organic aromatic esters that are either dialyl or alkyl aryl. Phthalates are almost exclusively human-manufactured chemicals; they are not found in nature except for small amounts in coal and crude oil (Health Canada, 2017). The two phthalate side chains can be identical or different, and chemical properties of the phthalates are determined by the alkyl, alkenyl, and aryl chains' polarity, length, and molecular weight (Patnaik, 2017). Phthalates have been broadly used in plastics dating back to the 1930s (Graham, 1973).

Figure 2.1: Basic Molecular Structure of Phthalates

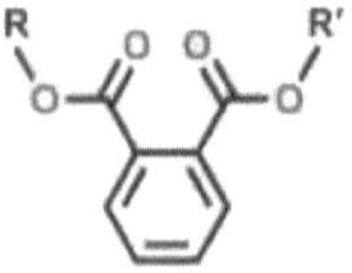

Figure from Crawford, C. B., & Quinn, B. (2017). 6 - The interactions of microplastics and chemical pollutants (C. B. Crawford & B. B. T.-M. P. Quinn, Eds.).p. 138.

Phthalates, which are liquid at room temperature in their pure form, have a high boiling point (Staples et al., 1997) and low water solubility (Ellington, 1999; Staples et al., 1997). In the environment, phthalates degrade under aerobic, anoxic or anaerobic conditions (Jianlong et al., 2000) Short chain phthalates typically degrade more rapidly than long chain phthalates (Jianlong et al., 2000; Liang et al., 2008). Phthalates are organic compounds that do not covalently bind in mixtures. Instead, when phthalates in chemical mixtures are heated, they form polar bonds that are relatively stable at room temperature (Graham, 1973). This makes them an ideal compound to impart flexibility in plastics, but also means they can leach, migrate, or evaporate to the environment under the right conditions (Heudorf et al., 2007).

Phthalates are often subdivided into two groups based on chemical structure and properties. The low molecular weight phthalate (LMWP) group typically includes phthalates with 1-4 carbon side chains, including di-n-butyl phthalate (DBP), diethyl phthalate (DEP), butyl benzyl phthalate (BBzP), dimethyl phthalate (DMP), and diisobutyl phthalate (DiBP) (Environment and Climate Change Canada & Health Canada, 2015a). The medium and high molecular weight phthalate group (HMWP), which have five or more carbon side chains, include chemicals such as dicyclohexyl phthalate (DCHP), diisononyl phthalate (DiNP) di-n-octyl phthalate (DnOP) and di (2-ethylhexyl) phthalate (DEHP) (Environment and Climate Change Canada & Health Canada, 2015a, 2015c). More recently, newer phthalates and replacement plasticizers have been developed to act as analogues for DEHP and other restricted phthalates. These "replacement" chemicals – including DiNCH (1, 2-Cyclohexane dicarboxylic acid diisononyl ester) and DEHTP (di-2-ethylhexyl terephthalate) – are used in a variety of consumer goods.

In biomonitoring studies, phthalate metabolites, as opposed to parent compounds, are typically measured in urine as a means to quantify exposure. In this study, metabolite names are used as opposed to the name of parent compounds to accurately represent what was measured in the biomonitoring component of this study.

Table 2.1 – Phthalate parent and metabolites measured in the MIREC Study

	Parent Phthalate	Phthalate Metabolites
	butyl benzyl phthalate (BBzP)	mono-benzyl phthalate (MBzP)
Low Molecular Weight[1]	Di-n-butyl phthalate (DnBP)	mono-n-butyl phthalate (MnBP)
	Diethyl phthalate (DEP)	mono-ethyl phthalate (MEP)
	Dimethyl phthalate (DMP)	mono-methyl phthalate (MMP)

Molecular Weight Group[1]	Parent compound	Metabolite
Intermediate Molecular Weight	Di-cyclo-hexyl phthalate (DCHP)	mono-cyclo-hexyl phthalate (MCHP)
High Molecular Weight	Di-iso-nonyl phthalate (DiNP)	mono-isononyl phthalate (MiNP)
	Di-n-octyl phthalate (DnOP)	mono-n-octyl phthalate (MnOP) mono-(3-carboxypropyl) phthalate (MCPP)
	di (2-ethylhexyl) phthalate (DEHP)	mono-(2-ethylhexyl) phthalate (MEHP); mono-(2-ethyl-5-oxo-he[...] (MEOHP); and mono-(2-ethyl-5-hydroxy-hexyl) phthalate [...]

[1] Molecular weight groups based on Health Canada's proposed sub-groups determined by structure-activity (Health Canada, October 2011, Group profile for phthalates https://www.canada.ca/en/health-canada/services/chemical-substances/substance-groupings-initiative/phthalate/group-profile.html)
Low Molecular weight: R-alkyl backbone with 3 or less carbon atoms
Intermediate Molecular Weight: R-alkyl backbone of 4-6 carbon atoms
High Molecular Weight: R-alkyl backbone with a benzyl group or 7 or more carbon atoms

2.1.2 Phthalates in Consumer Goods

Phthalates are commonly found in a wide variety of consumer goods, and several types of phthalates are frequently found in the same products. LMWPs are commonly used in personal care products (PCPs), such as lotions, perfumes, soaps, and moisturizers. In a random sample of personal care products (PCP) in Canada, Koniecki and colleagues (2011) found that at least one phthalate was present in about half of the products tested, with DEP, a fragrance stabilizer, detected in 103 of 252 products tested (41%). Other studies have shown that PCP use is highly associated with short chain phthalate metabolite (MEP, MiBP and MnBP) concentrations in urine in American women who were recently pregnant (Buckley et al., 2012; Parlett et al., 2013). Similar results have also been seen in pregnancy cohorts(Fisher et al., 2019; Just et al., 2010). The P4 study (Fisher et al., 2019) used serial spot urine samples during pregnancy to examine the association between urinary phthalate metabolites and recent use of a wide variety of consumer products, and determined that high users of PCPs had significantly elevated urinary MEP compared to the medium- and low-use categories.

HMWPs in Canada are commonly found in plastics, particularly polyvinyl chloride (PVC) and construction materials (Environment and Climate Change Canada & Health Canada, 2015a; Meek & Chan, 1994). A cross-sectional study using data from the SELMA cohort in Sweden found that the presence of vinyl flooring in the home was associated with higher levels of BBzP metabolites in pregnant women (Shu et al., 2019). Higher levels of BBzP have also been found in the house dust of homes with vinyl flooring as compared to homes without vinyl flooring (Philippat et al., 2015). Children who live in homes with PVC flooring have been found to have higher levels of urinary MBzP as compared to those who do not live in homes with PVC flooring (Bornehag et al., 2005; Carlstedt et al., 2013; Hammel et al., 2019).

Phthalates are not naturally occurring in foods, but food is one of the major sources of exposure to HMWPs. HMWPs and to some extent LMWPs, are somewhat lipophilic, meaning they accumulate in non-polar fats. Ingestion of high-fat foods, including meat and meat products (Page & Lacroix, 1995; Serrano et al., 2014; Xiang et al., 2020), cooking oïl (Husøy et al., 2019; Kiralan et al., 2019; Serrano et al., 2014; Xiang et al., 2020) high-fat dairy products (Husøy et al., 2019; Serrano et al., 2014) and fast food (Buckley et al., 2019; A. R. Zota et al., 2016) are highly associated with urinary HMWP metabolites. Exposure to HMWP in these food sources is likely caused by contamination via plastic food packaging or processing. A recent review concluded that overall, dietary predictors of phthalate exposures in pregnancy are the same as for individuals in the same age group who are not pregnant (Pacyga et al., 2019).

In Canada, the Phthalate Substance Grouping includes 14 compounds of interest (Environment and Climate Change Canada & Health Canada, 2015b), but over 25 types of phthalates are commonly found in a wide variety of consumer goods. Industry also continues to develop new phthalates and analogues to replace and improve existing chemicals. The prevalence of these chemicals in Canada is ubiquitous but asymmetrical: in 2012, over 10 million kilograms of DINP, DIDP, DUP, DEHP, D911P and DIUP were imported or manufactured in Canada, while fewer than 100kg per year of BCHP, CHIBP, DBzP, DMCHP, BIOP, DnHP and DPrP were imported or manufactured, and all other phthalates (15 measured) were between 10 000 to 1 000 000 kg/year (Environment Canada, 2017).

2.2 Human Exposure and Measurement

2.2.1 Routes of Exposure

Given the high prevalence of phthalates in the human environment, phthalates can enter the human body through several routes of exposure, including dermal absorption, ingestion, or inhalation (Schettler et al., 2006). The Chronic Hazard Advisory Panel (CHAP) on Phthalates, a working group from the United States convened to "study the effects on children's health of all phthalates and phthalate alternatives as used in children's toys and child care articles" (U.S. Consumer Product Safety Commission, 2014), concluded that all age groups are exposed to phthalates from a variety of sources, but direct ingestion (food, beverages, medication) accounts for the highest exposures (Lioy et al., 2015). As previously mentioned, food typically does not naturally contain phthalates, but phthalates may be present in food sources due to contamination or leaching. Absorption of phthalates occurs primarily through the use of PCPs, while inhalation occurs primarily in indoor environments, especially in the presence of vinyl flooring (Carlstedt et al., 2013; Just et al., 2015)

2.2.2 Exposure Measurement and Biomonitoring

Measuring human exposure to phthalates is a long-standing challenge (Jaeger & Rubin, 1970). Several methods can be used in population studies to determine exposure levels, including passive sampling, estimation based on questionnaires, and biomonitoring. Questionnaires or passive sampling typically provide more qualitative data on phthalate exposure, while biomonitoring is the gold standard for quantitative exposure classification (Calafat & Needham, 2009).

In studies where researchers are more interested in understanding the types of chemicals a person is

exposed to, as opposed to the quantitative exposure of each chemical, questionnaire methods can

provide some information on exposure (Fierens et al., 2012). Questionnaire methods are relatively

inexpensive to administer, but are prone to participant memory loss and self-report bias. For general

monitoring and identification of phthalate exposure, passive samplers are able to identify the presence

of a variety of chemicals. Recently, a novel technique using silicone wristbands as passive samplers has

been developed as a cheaper alternative to more laborious biomonitoring methods (O'Connell et al.,

2014). While this method does not measure the quantity of a chemical within the human body, it does

give researchers access to qualitative exposure data using a minimally invasive tool that is inexpensive

and easy to use. Passive sampling with silicone wristbands has been used in several recent phthalate

studies, including a study that found that phthalates are commonly identified in populations in Africa,

South America, and the United States, where DEHP, DnBP, DiNP, DEP, and DiBP were found on over half

of the wristbands used in all locations (O'Connell et al., 2014).

Biomarkers are the gold standard for measuring human exposure to phthalates (Needham et al., 2007).

Human biomonitoring is defined as the practice of measuring body burden using biological markers or

biomarkers from humans (Angerer et al., 2007). Biomarkers are commonly used in phthalate exposure

studies, but the chemical properties of phthalates pose some unique challenges in epidemiological

studies (Calafat & Needham, 2009; L E Johns et al., 2015).

Phthalates can be found in several biological matrices, including breast milk, serum, saliva, infant

meconium and amniotic fluid (T. Arbuckle et al., 2016; Calafat et al., 2015; M J Silva et al., 2004). Urine is

the most commonly used matrix for measuring body burden of phthalates because of the relatively non-

invasive nature of urine collection, and because of the enzymatic stability of phthalate metabolites in

urine, which decreases further breakdown (Samandar et al., 2009). Phthalate body burden is typically determined based on the concentration of monoesters and conjugated monoesters in urine, the by-products of phthalate metabolism in the body (Meeker et al., 2009). The half-life of phthalates in the human body is approximately six hours, although Koch (2005) reported only 75% excretion of DEHP metabolites in an adult male volunteer after 48 hours. Concentrations of phthalate metabolites in human samples are typically measured in a laboratory using tandem High Performance Liquid Chromatography / Mass Spectroscopy (HPLC-MS/MS) (Koch et al., 2003).

Single spot urine samples, or a urine sample taken at one time point, are routinely used to measure an individual's level of exposure to phthalates. Unfortunately, this type of exposure characterization may lead to misclassification of exposure because phthalates are cleared rapidly, and the concentrations of phthalates in urine vary depending on the route of phthalate exposure, time of the day that the sample is collected, and season of collection (Adibi et al., 2008; Aylward et al., 2017; Fisher et al., 2015). In general, phthalates measured in 24-hour urine samples are less prone to misclassification than a spot urine sample (Fisher et al., 2015; Perrier et al., 2016). Multiple urine samples, which are better than a single spot urine sample, are not always feasible in cohort studies because they are more laborious for the participants and more expensive to analyze. Despite some limitations, single spot urine samples are acceptable for use in epidemiological studies because they are prone to non-differential misclassification, meaning the results would likely be biased towards the null.

Using urine for biomonitoring is complicated by the risk of contamination during sample collection or analysis. Laboratory supplies or equipment used to collect or analyse urine samples are often made of phthalate-containing soft plastics, which can contaminate samples (Calafat & Needham, 2009; Reid et al., 2007). To avoid contamination, phthalate results are typically compared to a sample blank to account for processing contamination. Together, the rapid metabolism of phthalates and contamination both pose challenges for measuring exposure in human populations.

2.2.3 Phthalate levels in the general population

The general population is widely exposed to phthalates, but the type of phthalate exposures can vary over time. Globally, several large-scale surveys using urinary biomarkers have time trend measurements of phthalates in human populations. Regional trends in exposures to various phthalates indicate that government regulations, industry restrictions, and consumer behaviors affect the prevalence of phthalate exposures.

In the United States, urinary phthalate concentrations in the general population are monitored by a periodic random sample of the National Health and Nutrition Examination Survey (NHANES), a nationally representative longitudinal sample of the US population. Phthalate data collected from the last 20 years of NHANES suggest that in the United States, exposure to DEP, DnBP, BBzP, and DEHP decreased significantly between 2001 and 2014 (Environmental Protection Agency, 2017; A. Zota et al., 2014) During the same time period, however, concentrations of new "replacement" phthalates such as DiBP and DiNP metabolites increased (Zota et al., 2014). More recently, NHANES samples have been used to measure other replacement plasticizer metabolites such as DEHTP and DiNCH (Silva et al., 2019). Human exposure to these replacement chemicals will undoubtedly increase in the coming years as they become more prevalent in consumer goods (Rodríguez-Carmona et al., 2020; Silva et al., 2013).

Phthalate levels in the European Union follow similar trends to the United States. Similar to the United States, Europe has implemented limits on DEHP, DnBP, DiBP and BzBP in children's toys, clothing, and accessories. Several other phthalates (Diisopentyl phthalate [DIPP], Dipentyl phthalate [DPP], n-Pentylisopentyl phthalate [nPiPP], and Bis[2-Methoxyethyl] phthalate [DMEP]) have been deemed "toxic for reproduction," and will gradually be phased out of use (Tranfo et al., 2018). Several European cohort and cross-sectional studies show that exposure to regulated phthalates has been decreasing. A cross-sectional study of the general population in Italy indicates that DEHP, BBzP, DEP, and DnBP metabolites have decreased in European populations between 2011 and 2016 (Tranfo et al., 2018). Similarly, in a population of young Danish men, urinary metabolites of DiBP, DnBP, BBzP and DEHP were over 50% lower in 2017 compared to 2009 (Frederiksen et al., 2020). In Sweden, DEP, DnBP, and BBzP all decreased from 2009 to 2014 (Gyllenhammar et al., 2017). At the same time, exposure to "replacement" phthalates is increasing in Europe; DEHTP and DINCH increased in Denmark (Frederiksen et al., 2020); DiNCH increased in Sweden (Gyllenhammar et al., 2017; Shu et al., 2018) and in Germany (Kasper-Sonnenberg et al., 2019)

Several countries in Asia later adopted similar phthalate restrictions as America and Europe. In 2008, China restricted phthalates in food containers and other packaging materials (Hygienic Standards for Uses of Additives in Food Containers and Packaging Materials, 2008). This restriction may have resulted in decreased phthalate exposure; a recent cross-sectional study of house dust in one Chinese city found that median levels of phthalates found in the dust were similar to or lower than studies in European or North American populations prior to restriction (Q. Zhang et al., 2020). Similarly, in Korea, DEHP, DBP, and BBzP were all prohibited in food storage products (Korean Ministry of Government Legislation, 2010) which may be related to decreasing levels of phthalate exposure in the general population (Choi et al., 2017).

Trends in Asia vary by country. A study of healthy adult volunteers in six Asian countries (China, India, Japan, Kuwait, Malaysia and Vietnam) in 2010 found that participants in all countries except Kuwait had DEHP metabolite levels comparable with or lower than participants in the US NHANES survey, as well as lower MBzP exposure, and lower MEP exposure (except for Kuwait which was higher and India which was comparable) (Guo et al., 2011). Metabolites of MnBP and MiBP were significantly higher than in the US population, particularly in China and Kuwait. Chinese surveys consistently show that the majority of their phthalate exposure is to metabolites of MnBP and MiBP (Guo et al., 2011; Q. Zhang et al., 2020);

this differs from most studies in Europe and North America where MEP is typically the metabolite with the highest average concentration in biomonitoring studies.

Endocrine-disrupting chemicals are an emerging concern on the African continent; little is known, however, about human exposure to phthalates (Beltifa, Belaid, et al., 2018). A recent review found only three biomonitoring studies in African populations that investigated phthalate exposure and human health effects (Joubert et al., 2020) . Some small scale studies have found that phthalate plasticizers are present in personal care products at similar levels to those found in other international populations, including a study in Tunisia that found that DEP was present in personal care products (Beltifa, Belaid, et al., 2018). In the same study, DEHP was found in several samples, despite being banned in personal care products (Beltifa, Belaid, et al., 2018). Convenience samples of cheese from Tunisia were also found to be contaminated with several phthalates, including high levels of DBP and DEHP (Beltifa, et al., 2018b). In one study, phthalates were detected in over 50% of passive samplers worn by participants in several different locations in South Africa and Senegal (Dixon et al., 2019), and a small biomonitoring study in Egypt found that pre-pubescent girls were exposed to phthalates at a level comparable to participants in NHANES (Colicino et al., 2019).

The largest biomonitoring program in Canada is the Canadian Health Measures Survey (CHMS) directly measures a variety of health outcomes and exposures on a representative sample of approximately 5,000 Canadians (3 to 79 years of age) per cycle, excluding: persons living in the three territories; persons living on reserves and other Aboriginal settlements in the provinces; full-time members of the Canadian Forces; the institutionalized population and residents of certain remote regions (Statistics Canada, 2020). t CHMS measured 11 phthalate metabolites in urine for participants aged 6-49 between 2007-2009 (cycle 1) and for participants aged 3-79 between 2009-2011 (cycle 2) and 2016-2017 (cycle 5) (Health Canada, 2019). Mean exposure levels cannot be directly compared for all age groups among the

CHMS because of the difference in the age groups sampled in Cycle 1 as compared to cycle 2 and 5, but the study gives a good indication of population level exposures. Canadians were universally exposed to all of the phthalates measured, with few exceptions (MCHP, MMP, MiNP, MOP) (Health Canada, 2017). Similar to the United States, the geometric mean of phthalate metabolites measured in CHMS decreased over time for MEP, DnBP, MBzP, and the DEHP metabolites (Health Canada, 2019). The mean MCPP levels were similar in cycle 1 and cycle 2 of CHMS, but lower in cycle 5.

There are limited data currently available on Canadians' exposure to newer "replacement" plasticizers (Health Canada, 2019). Of the six metabolites measured, four were detected, but only in children. 3-5 Cyclohexane-1,2-dicarboxylic acid (CHDA), 3-5, cis-Cyclohexane-1,2-dicarboxylic mono carboxyisononyl ester (cis-cx-MINCH), Cyclohexane-1,2-dicarboxylic mono hydroxyisononyl ester (OH-MINCH), and Cyclohexane-1,2-dicarboxylic mono oxoisononyl ester (oxo-MINCH) were all detected above the limit of detection in more that 40% of child participants aged 3-5; OH-MINCH and oxo-MINCH were also detected in more that 40% of children aged 6-11.

2.3 Phthalates and Human Health

2.3.1 History of Phthalate-Related Human Health Concerns

Like many non-persistent chemicals, phthalates were initially not regulated due to their perceived safety based on animal toxicology studies and rapid metabolism in animal models (Autian, 1973; Shaffer et al., 1945). Phthalates were known to leach out of plastics, particularly in high-lipid environments (Graham, 1973). Early studies identified phthalates in tissue samples from patients who had received blood transfusions (Jaeger & Rubin, 1970) and in the lipid extracts of their blood (Marcel & Noel, 1970) but neither of these findings raised concerns at that time among toxicologists.

Toxicologists began to identify the endocrine-disrupting nature of phthalates in animal models during the 1980s and early 1990s (Gray & Gangolli, 1986; Laskey & Berman, 1993; Lloyd & Foster, 1988). Around the same time, the impacts of low-level chemical exposure and related human health effects due to endocrine disruption were beginning to emerge (Colborn et al., 1993). As a result, investigations into the human health effects of phthalates increased significantly in the following decades. Phthalates have now been linked to a wide variety of hormone-specific health outcomes in children, ranging from neurodevelopment to pubertal timing to changes in metabolic health (Meeker, 2012).

Phthalates and phthalate metabolites have been identified by international organizations such as the Endocrine Society (Diamanti-Kandarakis et al., 2009; Gore et al., 2015) and the World Health Organization (Bergman et al., 2012) as chemicals of potential concern because of their associations with endocrine disruption and related adverse health effects in humans, including impacts on male reproductive health, thyroid health, and metabolic health. In the United States, the *Consumer Product Safety Improvement Act* established an advisory panel (U.S. Consumer Product Safety Commission, 2014) to study the health effects of phthalates (U.S. Consumer Product Safety Commission, 2014). The current understanding of human health concerns related to phthalates is rooted in evidence of their endocrine-disrupting activity, and work continues to investigate these health effects, particularity during periods of sensitivity, such as early childhood, puberty, and during pregnancy.

2.3.2 Environmental exposures during early life – The MIREC Study

Understanding critical periods of human development, and the ways in which environmental exposures during these critical periods can disrupt development is an area of increasing research in environmental epidemiology (Braun & Gray, 2017). The Developmental Origins of Health and Disease (DOHaD) hypothesis was initially proposed by David Barker (1998) after observing the relationship between poor nutrition in mothers and disease outcomes in children. The idea that early life exposures may impact an individual throughout the lifespan has impacted environmental health research for over 20 years (Haugen et al., 2015). Over nearly the same time period, epidemiological studies that include exposure assessments have tried to incorporate more comprehensive exposure assessments in order to avoid misclassification of exposures, and more accurately identify the components of concern within these mixtures (Wild, 2005, 2012).. To identify links between developmental exposures and disease, it is important to prospectively examine the sum of exposures from the prenatal period onwards and catalogue whether and how those exposures are impact disease.

With this in mind, the Maternal-Infant Research on Environmental Chemicals (MIREC) Study, a pan-Canadian pregnancy cohort study, was launched in 2008 to investigate the impact of chemical exposures on pregnant women and their children (Arbuckle et al., 2013). From 2008-2011, women in their first trimester of pregnancy were recruited from 11 sites across Canada to participate in the cohort (Arbuckle et al., 2013). Women were eligible for inclusion if they were between ages 18-45, had no history of drug abuse, complicated pregnancy, or heart conditions. In total, 1983 mothers agreed to participate, and 1959 live births were recorded. While not nationally representative, MIREC represents one of the most comprehensive longitudinal cohorts in the world, and includes over 10 years of extensive information on both maternal biomonitoring, clinical and questionnaire data, and child biomonitoring, testing and questionnaire data.

MIREC is currently the largest Canadian mother-child cohort study with extensive data on phthalate exposure during early pregnancy. In MIREC, a single void urine sample was collected from each participant during the first trimester of pregnancy and analysed for a variety of non-persistent chemical compounds. Phthalates were measured using tandem MS-MS/HPLC. 11 phthalate metabolites were measured in the urine samples (Table 1). MEP, MnBP, and four DEHP metabolites were detected in over 95% of the urine samples provided, while MCHP, MMP, MiNP, and MOP were detected in less than 15%

of urine samples (Table 2.2) (T. Arbuckle et al., 2014). The median concentrations for MEP, MnBP,

MEHP, MEOHP and MEHHP each were comparable to the concentrations reported for all women aged

20-29 in the Canadian Health Measures Survey (Cycle 2, 2009-2011) (Health Canada, 2017).

Table 2.2: Concentration of phthalate metabolites (µg/L) measured in the first trimester urine samples in the MIREC study

Metabolite	D*n*BP MnBP	DEP MEP	BBzP MBzP	DMP MMP	DCHP MCHP	DiNP MiNP	DnOP MOP	MCPP	DEHP MEHP	MEOHP	MEHHP
MLE GM (95% CI)	11.59 (10.96–12.26)	32.02 (29.75–34.47)	5.20 (4.90–5.52)	ND	ND	ND	ND	0.86 (0.80–0.92)	2.24 (2.12–2.37)	6.39 (6.04–6.75)	9.16 (8.65–9.71)
K–M Median (95% CI)	12.00 (11.18–12.82)	28.00 (25.45–30.55)	5.20 (4.84–5.56)	NA	NA	NA	NA	0.92 (0.85–1.01)	2.20 (2.06–2.34)	6.50 (6.16–6.84)	9.40 (8.77–10.03)
95th percentile	69.65	530.00	41.65	10.0	0.38	NA	NA	9.26	15.00	41.00	65.65
Maximum	3100	13,000	420.00	1000	77.00	9.20	7.90	370.00	340.00	980.00	1200.00
% < LOD	0.28	0.17	0.67	85.35	92.23	98.49	97.82	17.84	2.35	0.45	0.95
N	1788	1788	1788	1788	1788	1788	1788	1788	1788	1788	1788
SG-standardized											
MLE GM (95% CI)	13.69 (13.15–14.24)	37.73 (35.37–40.24)	6.14 (5.86–6.43)	ND	ND	ND	ND	1.02 (0.97–1.08)	2.63 (2.52–2.74)	7.54 (7.24–7.84)	10.81 (10.36–11.28)
K–M median	13.00 (12.51–13.49)	30.95 (28.48–33.42)	5.70 (5.38–6.02)	NA	NA	NA	NA	0.95 (0.91–1.00)	2.38 (2.28–2.48)	6.93 (6.71–7.16)	9.88 (9.52–10.24)
95th percentile	50.78	486.91	37.41	21.67	0.88	NA	NA	7.60	12.21	30.94	50.35
Maximum	1831.82	20,800	342.73	541.67	47.67	5.46	4.89	186.33	260.00	733.57	1114.29
% < LOD	0.28	0.17	0.67	85.35	92.23	98.49	97.82	17.84	2.35	0.45	0.95
N	1785	1785	1785	1785	1785	1785	1785	1785	1785	1785	1785

LOD: limit of detection.
K–M median: Kaplan–Meier median, censored method.
MLE GM: maximum-likelihood estimated geometric mean, censored method.
ND: below limit of detection.
NA: not applicable due to high level of censoring
SG: Specific Gravity

Table modified from Arbuckle, T., Davis, K., Marro, L., Fisher, M., Legrand, M., LeBlanc, A., ... Group, M. S. (2014). Phthalate and bisphenol A exposure among pregnant women in Canada--results from the MIREC study. *Environment International, 68*, 55–65.

2.3.3 Epidemiological evidence for association between phthalates and neurodevelopment

Epidemiological concerns regarding phthalates stem from evidence in animal models that phthalate exposure is related to many hormone-dependent adverse outcomes. Exposure to phthalates, specifically DBP, DEHP, and BzBP, is linked to a group of outcomes known as the phthalate syndrome in rodents, with some evidence in rabbits (Foster, 2006). Phthalate syndrome occurs when phthalate exposure affects testosterone production at critical time points, resulting in a variety of adverse effects on sexual development, including atypical development of the vas deferens and seminal vesicles, external male genitalia, reduced anogenital distance, and cryptorchidism. This evidence of hormone disruption in animal models has lead to interest in other potential disruption in hormone-dependent processes in humans. Neurodevelopment is known to involve several hormone-dependent processes, and specific interest in the association between phthalate exposure and IQ scores began after associations were noted between prenatal phthalate metabolites levels and developmental scores in young children in cohorts in the United States (Whyatt et al., 2012) and South Korea (Y. Kim et al., 2011).

In human populations, epidemiological studies have examined the association between prenatal phthalate exposure and neurodevelopmental outcomes, including language development (Olesen et al., 2018), autistic traits (Oulhote et al., 2020; Testa et al., 2012), motor and neurodevelopmental scores (Doherty et al., 2017; Kim et al., 2018; Kim et al., 2011; Polanska, et al. 2014; Tellez-Rojo et al., 2013; Whyatt et al., 2012) and Attention Deficit Hyperactivity Disorder (ADHD) (Engel et al., 2018). Many aspects of intellectual function are impacted by hormone-dependent processes in-utero (Braun, 2017) which may make them susceptible to alterations in phthalate exposure. A recent review of phthalate exposure and neurodevelopment (Radke et al., 2020) found moderate evidence that BBzP was associated with motor ability score, but was inconclusive for other phthalates and phthalate metabolites.

Intellectual function measured using IQ scores is one way of measuring cognitive development; there are many others. While IQ scores are not a direct measurement of neurodevelopment, they are found to be associated with both demographic factors and brain development (Gale et al., 2006; Lange et al., 2010; Matthews et al., 2018; Reiss et al., 1996)A wide variety of studies have examined the association between prenatal phthalate exposure and IQ scores (Factor-Litvak et al., 2014; Huang et al., 2015; Hyland et al., 2019; Jankowska et al., 2019; Kim et al., 2017; Li et al., 2019; Nakiwala et al., 2018; Tanner et al., 2019). These studies used standard scales to measure IQ, such as the Wechsler Preschool and Primary Scale of Intelligence (WPPSI), which measures IQ in children between ages 2.5 and 7.5, and the Wechsler Intelligence Scale for Children (WISC), which measures IQ in children ages 6 to 17. These scales are validated for use on a wide variety of populations.

In a North American cohort, Factor-Litvak et al. (2014) found that third trimester urinary MBP and MiBP metabolites were inversely associated with Child IQ (WISC-IV) measured at age 7. In contrast, in the CHAMACOS cohort, Hyland and her colleagues found that the sum of all first trimester HMWPs (MBzP, MCPP, MCOP, and MCNP) were inversely associated with Full Scale IQ, and MEP concentrations were inversely associated with Working Memory IQ scores (Hyland et al., 2019). Li and colleagues reported that urinary MBzP at 16 weeks gestation was inversely associated with IQ in a mixture analysis of 26 environmental chemicals (Li et al., 2019). In a single-chemical analysis of the same cohort, MCPP measured at 16 weeks and 26 weeks gestation was inversely associated with IQ, though the association was not statistically significant (per 1-SD increase in concentration, 16 weeks β= -1.0 (-2.5, 0.6), 26 weeks β=-0.8 (-2.4, 0.7))(Li et al., 2019). The inconsistent findings across studies are further complicated by the fact that two of the studies examined each phthalate independently; only one study considered associations with chemical mixtures.

2.4.1 Mixture Methods – Background and History

In daily life, humans are exposed to a wide number of chemicals, and these chemical mixtures can pose significant challenges in environmental epidemiology. Complex mixture analysis in observational studies pose a variety of challenges for statistical analysis. Modelling techniques are limited in their ability to assess high dimensional chemical exposures (where the number of features is larger than the number of observations),, complex and correlated mixture exposure, low level and chronic exposure and also deal with relatively small risk differences when associations are observed (Pekkanen & Pearce, 2001). Finding ways to manage these limitations is critical in order to develop meaningful research conclusions and support policy development. Like many of fields of epidemiology, work in the field of mixture analysis is further complicated due to the interdisciplinary nature of the analysis, and often involves joint efforts by epidemiologists, toxicologists, risk assessors, and statisticians (Taylor et al., 2016).

While research into mixture analysis has been present in environmental epidemiology since the early 1990s (Carlin et al., 2013), there has been more formal emphasis on the field as a priority by international organizations in the last decade (National Institute of Environmental Health Sciences, 2012). Improvements in analysis software, including tools such as SAS, Stata, R, and SPSS have made analysis of large, complex data sets more feasible for a broader group of researchers. Several recent reviews and workshops (Braun et al., 2016; Carlin et al., 2013; Taylor et al., 2016) brought together interdisciplinary teams to assess the challenges and opportunities in the field of mixture analysis, and several new methods of analysis have been proposed and software developed as a result of this new focus (Bobb et al., 2015; Colicino et al., 2019; Keil et al., 2019; Tanner et al., 2019).

Methods developed for use in high dimensional mixture data fall into three general categories: dimension reduction, variable selection and observational grouping (Massimo Stafoggia et al., 2017).

Dimension reduction aims to deal with the problem of correlated exposures by reducing the number of exposures into a summary term or terms related to the outcome of interest. This can mean combining exposures and changing units, which can lead to issues in interpretability. Principal component analysis (Jolliffe, 2002)and Weighted Quantile Sum Regression (Carrico et al., 2015) are two examples of dimension reduction methods.

Variable selection methods typically involve identifying the most useful set of exposures for investigating an association, and ignoring other variables. Variable selection differs from dimension reduction because the method does not create a summary of all exposures, but selects which exposures are important to consider. Lasso (Tibshirani, 1996), elastic net (Zou & Hastie, 2005), and Bayesian kernel machine regression (BKMR) (Bobb et al., 2015)are types of variable selection methods.

The observational grouping methods assemble individuals with similar exposure profiles. This is distinct from dimension reduction, which creates new summary variables from exposures; observational grouping assigns a categorical variable to the individual based on similar exposure, and then uses this categorical variable to examine the association with an outcome. K-means clustering is a common example of observational grouping methods (Steinley, 2006).

Within the three general categories of statistical tools for high dimensional data, most methods are designed to investigate linear associations, or to incorporate regularization and penalties as a way to deal with non-linear association. Some additional tools for complex mixture analysis, including quantile g-computation (Keil et al., 2020) and BKMR (Bobb et al., 2015), were designed to handle non-linear associations in mixture analysis and may be useful tools when there is not an assumption of linearity. However, methods designed to deal with non-linear associations are more computationally intensive, and require complex interpretation strategies.

All statistical tools designed to assess mixtures have their advantages and weaknesses; there is no single method that is optimal for every mixture investigation(Braun & Gray, 2017; Tanner et al., 2020). Within the categories of statistical tools, several options exist for investigation (Lazarevic et al., 2019; Stafoggia et al., 2017), and the choice of tool typically depends on the investigator's familiarity with each technique. The various analytic tools available to study chemical mixtures makes it difficult to compare results across studies because various techniques may amplify bias in certain scenarios and result in incomparable estimates.

2.4.2 Correlated Phthalate Exposure

Exposure assessment studies have found that human exposure to various phthalates are highly correlated in urine (Frederiksen et al., 2010; Kalloo et al., 2018) due to the common routes of exposure. This correlation is problematic when investigating associations using linear modeling strategies due to the collinear nature of the phthalates. Collinearity in modeling means that one exposure is also highly predictive of another exposure, and if the two exposures are placed in the same model, the resulting estimates can be highly skewed and unreliable.

As expected, phthalates metabolites in the MIREC cohort showed various levels of correlation (Figure 2.2). DEHP metabolites (MEHP, MEHHP and MEOHP) showed near perfect correlation. Similarly, the long-chain phthalate metabolites, which have similar routes of exposure, were all highly correlated. Only MEP, a short chain phthalate metabolite, showed weak correlations to other chemicals.

Figure 2.2: Pearson Correlation coefficients for phthalate metabolites measured in the MIREC Study in first trimester urine samples

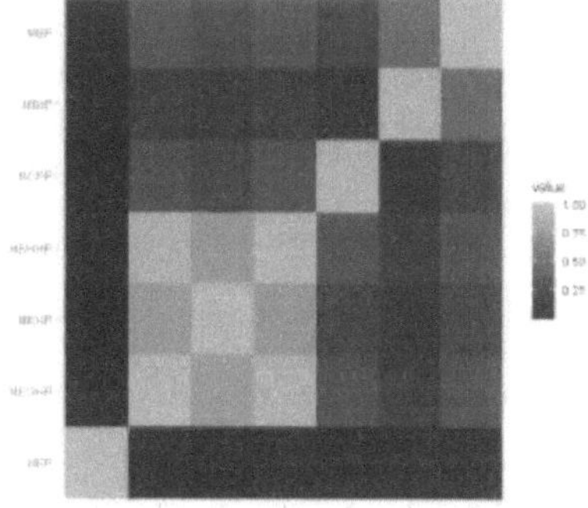

2.4.3 Weighted Quantile Sum Regression for Correlated Chemical Analysis

The Weighted Quantile Sum (WQS) regression method was initially developed and published in 2015

(Carrico et al., 2015). The method develops an index that summarizes the direction of association

between a mixture of exposures and an outcome of interest, and also identifies the components of the

mixture that most contribute to the association (Equation 2.1).

$$g(\mu_i) = \beta_0 + \beta_1 \left(\sum_{j=1}^{c} w_j q_{ji} \right) + z_i' \varphi$$

Equation 2.1: Weighted Quantile Sum Index (Carrico et al., 2015)

Each component in the mixture ($j=1$ to n), is binned into a quantile (q) of exposure for each individual (i);

and the weight (w_j) for each component (j) is then estimated using a trust-means regression algorithm.

The sum total of the weights of all mixture components will sum to 1. The equation can also incorporate

covariates and confounders (z) and their regression coefficients (φ), which are determined prior to

estimating the chemical weights in the index.

While the WQS method is computationally intensive, it can be summarized in four steps. In the first step, the available data are split into two sets: the first for training, and the second for validation. This method is commonly used in machine learning algorithms to test the stability of estimates (Williams, 2017). In the second step, the training data set is used to estimate the contribution, or weight, of each chemical in the mixture to the outcome, while also considering potential confounders. The second step utilizes a non-linear trust means regression algorithm, which simultaneously estimates the regression coefficients (β_1) and the weights of the components (c). The weight (w) of each component (c) is between 0 and 1, and the total sum of all of the weights of all the components is summed to 1 ($w_{c1} + w_{c2}$ ...$+w_{cn} = 1$). The direction of the regression coefficient is typically constrained to one direction (positive or negative) based on *a priori* knowledge of the association between the mixture and the outcome of interest. In the third step, step two is repeated multiple times using bootstrapping, or resampling with replacement. The bootstrapped sample is a randomly chosen sample (sampling with replacement) of individuals (i) from the training set. This application of bootstrapping allows for repetition, and an estimation of uncertainty associated with the regression coefficients and weights. The bootstrapped sample can be used to repeat step two, the estimation of the regression coefficients and weights of the component. This resampling step is repeated, typically between 100 to 1000 times. The final weight estimates are based on the average of the bootstrapped samples. Finally, in step 4, the significance of the WQS index is estimated by testing the β_1 estimate using the validation data set.

The WQS method has limitations, such as an inability to account for interactions in the mixture term and an unclear ability to handle highly correlated chemicals that are weakly associated with the outcome. WQS also has the potential to over fit random noise in the training data so that the estimates may be unreliable in general samples. Recently, methods have been developed to extend the WQS method to deal with a variety of scenarios, including chemical interactions within the mixture (M. Lee et al., 2019), a distributed lag framework to consider data from multiple time points (Bello et al., 2017), and an extension of the method to avoid the assumption of directional homogeneity (Keil et al, 2020). To deal with the potential of overfitting the model and developing unstable estimates, Tanner and colleagues suggested including repeated partitions of the data sets (Tanner et al., 2019). Essentially, they suggest performing repeated splits of the data into training and validation sets (step 1 as described above) to avoid the chance of an unrepresentative partition. The repeated partition method is essentially multiple iterations of the entire WQS procedure, steps 1-4 described above, with the final estimate of the relationship represented by the mean chemical weights and beta estimates. This repetition method is computationally more intensive than base WQS, but allows the user to estimate the mean chemical weights and beta estimates for a higher degree of stability. Based on investigative studies (Tanner, Bornehag, et al., 2019; Tanner, Hallerbäck, et al., 2019), the repeated partition method appears to attenuate the estimate of association towards the null as compared to the single partition method.

Several methods can be used to investigate correlated phthalates mixtures, but WQS has become increasingly popular (Daniel et al., 2020; Li et al., 2019; Romano et al., 2018; Shu et al., 2019; Tanner et al., 2019; Zhang et al., 2019) WQS is advantageous over many other mixture analysis methods because it is user-friendly, easily interpretable, and was designed explicitly for investigating correlated chemicals. The inclusion of the repeated partition extension generates more stable WQS estimates in smaller epidemiological samples, and allows for more extensive characterization of uncertainty in the estimates of chemicals of concern (Tanner, Bornehag, et al., 2019).

Chapter 3 – The MIREC Study

3.1 Background

Biomonitoring studies have been a major focus of environmental epidemiological studies since the

field's beginnings. In the late 1990s and early 2000s, there was an increase in research on environmental

exposures to metals and other toxicants, especially in vulnerable populations, such as pregnant women

and children. Specifically, there were concerns that exposure to chemicals like lead, PCBs, and mercury

in children, even at low levels of exposures, were associated with neurotoxicity and widespread harm

(Grandjean & Landrigan, 2006; Jacobson & Jacobson, 1996; Bruce P Lanphear et al., 2005; Ribas-Fitó et

al., 2001). Around the same time, growing evidence began to suggest that mercury may pass through

the placenta and impact neurodevelopment in a developing fetus (Grandjean et al., 1997) . In the early

2000s, there were no active biomonitoring studies of pregnant women or young children in Canada; in

fact, the most recent national study in Canada that examined environmental exposures had been

completed in the 1970s (Health and Welfare Canada and Statistics Canada, 1981). From a public policy

and public health perspective, more data were needed to assess health risks of current levels of metals

and other chemicals to which the general population was exposed.

The Maternal-Infant Research on Environmental Chemicals (MIREC) Study was designed to fill this

knowledge gap. MIREC was initiated as a pan-Canadian study to provide data on the environmental

exposures of pregnant women and children in Canada in non-occupational environments. MIREC was

also designed to provide additional data to what were collected in the CHMS. In Canada, levels of

chemical exposure have been collected through a nationally representative cyclical survey (the Canadian

Health Measures Survey (CHMS)) since 2007. The CHMS includes anthropomorphic measurements,

questionnaires for behavioural measurement, as well as blood measurements of health indicators (such

as environmental exposures and metabolic indicators) and urine measurements (environmental exposures, nutritional markers). However, the CHMS was not designed to provide data on pregnant women and young children (age range: 3-79 years). In addition, the CHMS is a cross-sectional survey, and does not provide the type of longitudinal information that is required for the dynamic biological processes of pregnancy.

3.2 Study Design

The Maternal-Infant Research on Environmental Chemicals (MIREC) study is an observational epidemiological study, designed as a multi-centre prospective cohort study. The MIREC Study built on pre-existing infrastructure of Canadian study sites (Figure 3.1) that were originally part of the International Trial of Antioxidants for the Prevention of Preeclampsia (INTAPP Study). Because of the unique developmental periods during pregnancy, MIREC was designed as a prospective cohort study, which allows researchers to monitor environmental exposures over time. This study design is particularly beneficial for pregnancy as It allow for repeated measures over windows of susceptibility for environmental exposures in the developmental process.

MIREC investigates several primary objectives, including assessing if exposure to maternal concentrations of metals such as lead are related to adverse pregnancy and birth outcomes, such as gestational hypertension, and reduced fetal growth. The MIREC study was also designed to provide extensive data on exposure to priority environmental chemicals during pregnancy across Canada, which was not a group that was studied in other large biomonitoring studies or surveys such as the CHMS. Other goals of the MIREC study were to collect genetic polymorphism data in order to investigate specific SNPs (single nucleotide polymorphisms) of interest related to chemical metabolism and known developmental pathways.

MIREC recruited participants from 11 study sites in 10 cities (Figure 3.1). Participants in the study were

recruited in their first trimester of pregnancy (6-13 weeks gestation) from participating research sites.

All research staff involved in the MIREC study were trained to ensure similar procedures for recruitment,

informed consent acquisition, and sample collection, thus reducing potential bias. Participation in the

MIREC study was voluntary, and informed consent was obtained from all participants. All protocols and

survey instruments were approved by Research Ethics Boards (REB) at Health Canada, as well as at every

participating research site. MIREC excluded participants based on known fetal abnormalities, known

major medical complications, and known drug use.

Figure 3.1 – Participating site locations and participants recruited for the MIREC, MIREC-ID, MIREC-CD3, and MIREC-CD Plus studies. Figure from https://www.mirec-canada.ca/en/about/some-facts-and-figures/

Site Number	City	Participating institutions	MIREC	MIREC-ID*	MIREC-CD3	MIREC-CD Plus		
					MSAQ**	MSAQ**	Bio-monitoring	Neuro-development
01	Vancouver	BC Children's and Women's Health Centre	162	34	45	38	83	55
02	Edmonton	University of Alberta	20		6			
03	Winnipeg	St. Boniface General Hospital/ Health Sciences Centre	90		23			
04	Toronto	Mount Sinai Hospital/ Sunnybrook Health Sciences Centre	325			70	108	72
05	Hamilton	McMaster University	275	114	47	75	112	85
06	Sudbury	Sudbury Regional Hospital	130		30			
07	Kingston	Kingston General Hospital	255	115	21	120	122	126
08	Ottawa	The Ottawa Hospital	119	12	50			
09	Montreal	CHU Sainte-Justine	300	127	58	115	175	145
10	Montreal	Jewish General Hospital	25	3	5	8	11	9
11	Halifax	IWK Health Centre	300	120	85	100	192	118
		Total	2001	525	370	526	803	610

*Participants who completed at least one of the two visits in MIREC-ID (visits at birth & at 6 months)

** MSAQ (Mother self-administered questionnaire) was completed as part of MIREC-CD3 or CD Plus, the total number of MSAQ is 896

The MIREC Study is a particularly robust research study because of the extensive data that were

collected throughout the study. MIREC involved study visits throughout the duration of the pregnancy and the early post-partum period. During these visits, participants provided biospecimen samples of blood, urine, and other biological samples (See appendix 2). Data from questionnaires were also collected for both parents, and medical chart review was completed for additional details. Several of these measures were repeated in different trimesters and soon after birth, providing longitudinal data. Meconium samples and cord blood were also collected from participating infants at delivery and milk from their mothers at follow up visits 2-10 weeks post delivery.

One unique feature of the MIREC study design was the inclusion of a biobank. Participants who agreed to inclusion in the biobank portion of the project agreed to storage of biological samples for future research on the health of pregnant women and their children for the duration of the project, or 30 years. As a result of this design, the MIREC biobank is a continued data source for researchers to answer new research questions, and for extended analysis with improved analytical methods.

The MIREC Study was funded through a variety of government agencies, including the Health Canada – Safe Environments Programme, Tobacco Control Programme, Product Safety Programme, Food Directorate, and Chemicals Management Plan, as well as the Ontario Ministry of the Environment, and Canadian Institutes of Health Research.

3. 3 Study Participants

8716 women were initially invited to participate in MIREC, of which 5108 met the eligibility criteria, and 2001 were enrolled in the study (Figure 3.2). Of those initially enrolled, there were 18 complete withdrawals (all biospecimens and questionnaires destroyed), as well as 56 partial withdrawals. Information is also missing for 48 participants lost to follow up and 14 participants who moved outside of any study sites (Arbuckle et al., 2013).

Figure 3.2 – MIREC Participant recruitment and participant chart. From Arbuckle et al, 2013

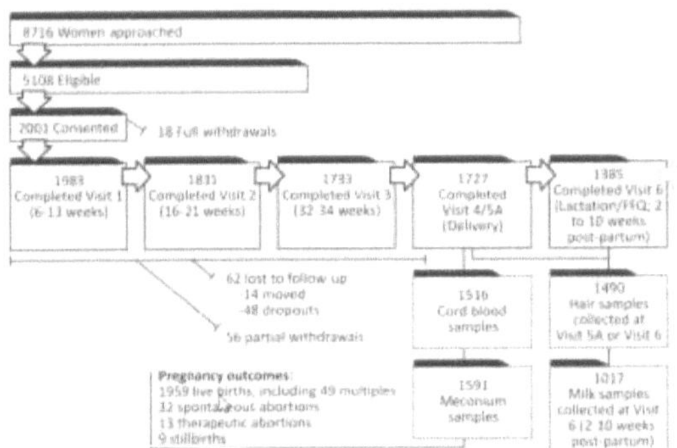

MIREC was not a population-based study due to the difficulties in recruiting representative samples in pregnancy cohorts. In an analysis of the cohort, the MIREC study team found that compared to Canadian women who gave birth in 2009, MIREC participants (recruited between 2008-2011) were found to be less likely to be a smoker than the general population, but more likely to be older. As compared to women who gave birth in 2009, MIREC mothers also had higher education levels, were more likely to be married, and were more likely to be born in Canada (Arbuckle et al., 2013). These differences, as compared to the general population of women who gave birth at the same time, may limit the generalizability of the MIREC results.

Data from the MIREC cohort remain the largest source of data on prenatal chemical exposure in Canada, and has been used in over 80 scientific publications (MIREC Research Webpage, 2021). MIREC represents an excellent data source for epidemiological study in that it utilizes questionnaires, in addition to biological samples, and chart review at several points throughout pregnancy and in the early post-partum period for the mother. With the success of the MIREC study, researchers were able to secure funding for several follow-up studies, allowing for continued longitudinal research with MIREC participants.

3.4 MIREC Biobank and Follow-up studies

Following the success of the MIREC cohort, there was strong enthusiasm for a continuation of the cohort. By the late 2000s, a growing body of research suggested that a wide variety of diseases were linked to environmental factors; a 2008 study (Boyd & Genuis, 2008) estimated that environmental exposures were associated with between 3.6 billion and 9.1 billion dollars in annual disease-related costs in Canada. With this continued concern over the link between environmental exposure and human health, several MIREC follow-up studies were established in order to continue high-quality research on the cohort. Each follow-up study focussed on their own individual research questions, and the sample size for participants varied based on resource restriction and participant retention. As of 2021, MIREC follow-up studies are still ongoing to continue to address important research questions.

3.4.1 MIREC-ID

MIREC-ID was established as the first MIREC follow up study in 2009. Participants at seven of the participating research sites (N=525) participated in study visits, immediately after birth and at 6 months of age. The focus of the MIREC-ID study was to investigate several anthropomorphic and developmental measures in participant infants, including early sexual development, behaviour, and heart rate variability.

3.4.2 MIREC-CD3

The MIREC-CD3 (Child development at age 3) study began in 2012 with 370 participants from all study sites except Toronto. The study was created to investigate the association between prenatal environmental exposures and neurodevelopmental as a focus of MIREC was on endocrine disruption; interest in neurological endpoints continued in MIREC follow-up studies. Specifically, the self-administered Behavioral Assessment System for Children-2 (BASC-2) and Behavior Rating Inventory of Executive Function (BRIEF-P) were used as measures of neurodevelopment.

3. 4.3 MIREC-CD Plus

MIREC-CD Plus, was established in 2013 with the specific research objectives of continuing biomonitoring studies for young children, and continuing to collect information on neurodevelopment for children at age 3 years. The CD Plus follow-up study recruited from the six MIREC sites with the largest MIREC participant population (Vancouver, Toronto, Hamilton, Montreal, Kingston and Halifax). The study focussed on biological specimen collection for child participants and neurodevelopmental testing, including tests investigating the children's behaviour, cognitive abilities, executive function, and language skills. The MIREC-CD Plus study aimed to enrol 800 children; 803 children participated in the biological specimen arm of the study and 610 participated in the neurodevelopment arm of the study.

The neurodevelopmental portion of the MIREC-CD Plus study included several self-administered and researcher-administered tests on behaviour and neurodevelopment. To evaluate behavioural outcomes, the Behavioral Assessment System for Children-2 (BASC-II) and Social Responsiveness Scale, Second Edition (SRS-2) were used; both of these assessments were self-completed by the MIREC mothers to report on their child's behaviour. To evaluate neurodevelopment and executive function, the Wechsler Preschool and Primary Scale of Intelligence – 3rd Edition (WPPSI-III), and NEPSY-II (A Developmental NEuroPSYchological Assessment) were administered by trained evaluators. The Behavior Rating Inventory of Executive Function - Preschool Version (BRIEF-P) was also completed by MIREC mothers to report different aspects of executive functioning.

3.4.4 MIREC-ENDO

The most recent MIREC study, the MIREC-ENDO (Endocrine) study, is currently in progress. MIREC-ENDO aims to investigate the association between prenatal environmental exposures and several endocrine – dependent processes during puberty, a time of major hormonal change and development. The MIREC-ENDO study is planned to occur over several phases, each of which will focus on different primary research questions. In MIREC-ENDO – Phase I, the main research question will investigate the association between prenatal environmental exposures and the onset of puberty. MIREC-ENDO also aims to investigate the association between prenatal environmental exposures and metabolic function in children and young teens and their mothers.

MIREC-ENDO phases two and three will begin in the next few years, to continue to gather information on adolescent development through puberty. The MIREC-ENDO cohort is uniquely situated as a natural experiment in a cohort that is reaching adulthood as Cannabis is legalized in Canada. The main goals of MIREC-Endo Phase II will be to investigate the association between current and previous cannabis and vaping use and child health. MIREC-ENDO Phase two will also assess the association of stress at different developmental windows and the timing of puberty in MIREC children.

3.5 MIREC Data in this study

In this book, data from MIREC, MIREC-ID, (breastfeeding status), and MIREC-CD Plus study (WPPSI IQ Scores, HOME Scores) are used for exposure, outcome, and covariate data. The significant amount of covariate data allowed us to investigate potential confounding, and provide a more complete analysis of the research questions.

Chapter 4 - Manuscript for publication: Individual and Cumulative Effects of a Mixture of Phthalates and Children's Intellectual Abilities: The MIREC Study

Background

Phthalates are a group of common chemicals found in consumer goods. Exposure to these chemicals is often highly correlated due to similar routes of exposure. Previous studies in animals and humans have suggested that phthalates act as endocrine disruptors. Epidemiological studies have identified that phthalates may interfere with normal fetal development, including neurodevelopment, but results are inconsistent and often only consider single chemical exposure.

Objectives

To examine potential associations between gestational (first trimester) phthalate exposure and neurodevelopment as measured using IQ scores at age 3-4 in the Maternal-Infant Research on Environmental Chemicals (MIREC) study.

Methods

Data from 607 mother-child pairs from MIREC were used in single-chemical linear regression models to assess the association between individual phthalates and Intelligence Quotient (IQ) Scores (*The Wechsler Preschool and Primary Scale of Intelligence – 3rd Edition*). We used Weighted Quantile Sum Regression (WQS) models to assess the association between urinary phthalate metabolites measured in the first trimester of pregnancy and IQ scores. Models were adjusted for study site, mother's education level, and mother's smoking status during pregnancy.

Results

Mono (3-carboxypropyl) phthalate was inversely associated with Full Scale IQ (-0.8, 95% CI: -1.5, 0.0) and Performance IQ (-1.0, 95% CI: -1.7, -0.1) in children in adjusted single chemical models. Higher exposure to a mixture of phthalate metabolites was inversely associated with FSIQ and PIQ, with mono (3-carboxypropyl) phthalate, mono-n-butyl phthalate, the sum of di (2-ethylhexyl) phthalate metabolites, and mono-ethyl phthalate identified as the main contributors in the FSIQ mixture (42%, 29%, 14% and 10%, respectively), and mono (3-carboxypropyl) phthalate, mono-n-butyl phthalate, and the sum of di (2-ethylhexyl) phthalate metabolites identified as the main contributors in the PIQ mixture (54%, 22%, and 14%, respectively)

Discussion

Gestational mono (3-carboxypropyl) phthalate may be associated with a decrease in IQ scores in children at age 3. In a mixture of phthalates, Mono (3-carboxypropyl) phthalate, mono-n-butyl phthalate, the sum of di (2-ethylhexyl) phthalate metabolites, and mono-ethyl phthalate metabolites may all contribute to lower IQ scores. Further research is needed to investigate the impact of prenatal exposure to phthalate mixtures.

Keywords
Phthalates, intellectual abilities, pregnancy, Weighted Quantile Sum Regression

Introduction

Phthalates are a large group of endocrine-disrupting chemicals. This group of chemical plasticizers and solvents includes chemicals that are diesters of benzenedicarboxylic acid with varying side chain length and structure (National Research Council (US) Committee on the Health Risks of Phthalates., 2008). Humans are frequently exposed to phthalates in daily life, where the chemicals are found in a variety of consumer goods, including building materials, food packaging materials, cosmetics, children's toys, and vinyl flooring (Schettler et al., 2006). Biomonitoring studies in Canada indicate near universal exposure to phthalates (Canada, 2013), including in susceptible populations, such as pregnant women and young children (T. Arbuckle et al., 2014, 2016). Phthalates are readily metabolized in the body (Koch et al., 2005; Meeker et al., 2009) and exposure to these chemicals is often highly correlated due to common sources of exposure (Saravanabhavan et al., 2013; Serrano et al., 2014).

Several studies have examined the associations between early-life phthalate exposure and a variety of adverse neurodevelopmental outcomes, including language development (Olesen et al., 2018), autistic traits (Oulhote et al., 2020; Testa et al., 2012), Intelligence Quotient (IQ) (Cho et al., 2010; Factor-Litvak et al., 2014; Huang et al., 2015; Li et al., 2019; Nakiwala et al., 2018), motor and neurodevelopmental scores (Doherty et al., 2017; Kim et al., 2018; Kim et al., 2011; Polanska et al., 2014; Téllez-Rojo et al., 2013; Whyatt et al., 2012), and Attention Deficit Hyperactivity Disorder (ADHD) (Engel et al., 2018).

Studies have reported inconsistent results of the association between urinary phthalate concentrations in pregnant women and IQ deficits in their children. Some cohort studies reported that gestational phthalate exposure was associated with lower child IQ (Cho et al., 2010; Factor-Litvak et al., 2014; Kim et al., 2018), while others reported no association (Huang et al., 2015; Kim et al., 2017; Polanska et al., 2014), or even positive associations (Nakiwala et al., 2018). However, previous pregnancy cohort studies have failed to account for the potential mixture effect of phthalate exposures, and inconsistently

identify metabolites of interest associated with the outcome of interest.

Phthalates pose a specific challenge in toxicological and epidemiological studies due to their non-

persistent nature and humans' highly correlated patterns of exposures in daily life. Analysing a mixture

of exposures using traditional environmental epidemiological methods is particularly challenging due to

the collinear nature of the data (Braun et al., 2016).

In this study, we investigated the association between phthalate exposures in the first trimester of

pregnancy and child IQ in the Maternal-Infant Research on Environmental Chemicals (MIREC) cohort. As

urinary phthalate concentrations in the MIREC cohort were correlated (Figure S4.5), this this study

considered individual effects of phthalate metabolites, and the cumulative effect of exposure to

multiple phthalate metabolites.

Methods
Study Population

Participants in this study were part of the Maternal-Infant Research on Environmental Chemicals

(MIREC) Study, and the follow up MIREC-CD Plus Study. This population has been described in detail

elsewhere (Arbuckle et al., 2013), but briefly, the MIREC study recruited pregnant women from 10 cities

across Canada during their first trimester of pregnancy (<14 weeks). Inclusion criteria included ability to

communicate in English or French, participant age 18 or older, <14 weeks gestation, willingness to

provide a sample of cord blood, and expectation of delivery in a hospital. Exclusion criteria included:

known fetal abnormalities with current pregnancy, illicit drug use, and known pre-existing maternal

health condition (kidney disease; epilepsy; any collagen disease including lupus, erythematous, and

scleroderma; active and chronic liver disease (hepatitis); heart disease; serious pulmonary disease;

cancer; haematological disorder (except anaemia or thrombophilia); threatened spontaneous abortion.

8716 women were asked to participate in the study, 5108 were eligible, and 2001 completed consent between 2008-2011. There were 18 complete withdrawals after recruitment,), resulting in 1983 participants with baseline data. Research Ethics Board Approval was obtained from all study sites and Health Canada, and informed consent was collected from all participants.

A subset of the MIREC participants (n=803) were enrolled in a follow-up MIREC-CD Plus study when participating children were up to 5 years old. The primary goal of the MIREC-CD Plus study was to evaluate associations between prenatal exposure to various chemicals and child growth and neurodevelopment. The MIREC CD-Plus study recruited from 6 MIREC sites (Vancouver, Toronto, Hamilton, Kingston, Montreal, and Halifax) and was completed in March 2015. 610 participants in the MIRE-CD Plus study completed IQ testing, n=2 were excluded due to age >4 years at testing, resulting in 608 participants with IQ scores collected at age 3-4 years. Of the 608 children with IQ scores, gestational phthalate data were missing for 45 due to lab error or insufficient urine volume for testing.

This study included 563 mother-infant pairs who had data on phthalate exposure during pregnancy from the MIREC study, and IQ scores for the children from the MIREC-CD Plus Study. The MIREC cohort included singleton and multiple births, but this analysis was restricted to singleton births.

Cognitive Outcomes

The Wechsler Preschool and Primary Scale of Intelligence – 3rd Edition (WPPSI-III) was administered to all MIREC-CD Plus child participants who were between 3 and 4 years of age at the time of testing. The test was completed at the participants' homes, and scored based on Canadian norms. Testing was completed by a trained evaluator at each site in the child's primary language (English or French). The evaluators had no knowledge of the child's exposure. The results of the 5 subtests (Receptive Vocabulary, Information, Block Design, Object Assembly, and Picture Naming) were used to determine

Verbal IQ (VIQ), Performance IQ (PIQ), and Full Scale IQ (FSIQ) scores. The population mean score for these tests is 100 (SD: 15).

The WPPSI-III is broadly used by psychology professionals, and shows high correlation to other intelligence tests including the Wechsler Intelligence Scale for Children (WISC) and the Bayley Scales of Infant Development (BSID). The WPPSI is validated on a variety of populations and is known to be reliable in the age group tested (Wechsler, 2002).

Laboratory Analysis

Participants provided a single spot urine samples during the first trimester of pregnancy. Time of day of collection and season of collection varied by sample. Within two hours of collection, urine samples were aliquoted into 30ml Nalgene tubes, frozen to -20C and shipped for processing. Prior to analysis, samples were stored at -20C in order to preserve metabolite stability (Samandar et al., 2009). Analyses for 11 phthalate metabolites (mono-n-butyl phthalate (MnBP); mono-ethyl phthalate (MEP); mono-benzyl phthalate (MBzP); mono-methyl phthalate (MMP); mono-cyclo-hexyl phthalate (MCHP); mono-isononyl phthalate (MiNP); mono-n-octyl phthalate (MnOP); mono-(3-carboxypropyl) phthalate (MCPP); mono-(2-ethylhexyl) phthalate (MEHP); mono-(2-ethyl-5-oxo-hexyl) phthalate (MEOHP); and mono-(2-ethyl-5-hydroxy-hexyl) phthalate (MEHHP)) was completed at Centre de Toxicologie du Québec, Institut national de Santé Publique du Québec (INSPQ) using LC–MS/MS with an Ultra Performance Liquid Chromatography (UPLC) Acquity (Waters; Milford, Massachusetts, USA) coupled with a tandem mass spectrometer Quattro Premier XE (Waters; Milford, Massachusetts, USA). We used field blanks to assess potential contamination from storage materials. We measured specific gravity in all urine samples using a refractometer (UG-1, Atago # 3461, Atago U.S.A. Inc., Bellevue, WA) in order to account for hydration status and urine dilution.

We chose metabolites with greater than 70% of samples above the limit of detection (LOD) for this analysis (MEP, MEOHP, MEHP, MEHHP, MCPP, MBzP, and MnBP). We investigated the significance of the exposure to DEHP by determining the molar sum concentration of all DEHP metabolites (MEOHP, MEHP, and MEHHP). DEHP was considered as a single exposure due to the high correlation of the metabolites (>85%).

MIREC included measures for several additional chemicals; cord blood lead and OP pesticide levels were included in this study. To measure blood lead levels, the INSPQ lab analyzed cord whole blood using Inductively-Coupled Plasma Mass Spectrometry (ICP-MS, PerkinElmer ELAN) at INSPQ. The lab also assessed OP pesticides in urine spot samples collected during the first trimester of pregnancy using GC-MS/MS (Agilent 7683; tandem mass detector, Waters Quattro Micro GC).

Covariates and Confounders

MIREC participants completed detailed questionnaires during the first trimester to collect demographic information, including age, population group (White, Other), household income, education level, medical history, and smoking status. Home Observation Measurement of Environment (HOME) Score (B. Caldwell & Bradley, 1984), a measure of the quantity and quality of caregiving in a child's life, was assessed by the same evaluator who completed the child's WPPSI-III evaluation as part of the MIREC-CD Plus evaluation.

We selected potential confounders identified in a literature review and drew a directed acyclic graph (DAG) (Greenland et al., 1999; Hernán et al., 2004) (Figure S4.1). Study site, age at assessment, maternal level of exposure to the three OP metabolites, maternal age at delivery, pre-pregnancy body mass index (BMI), mother's education level, household income, and mother's smoking status during pregnancy were identified as the minimal set of confounders. Cord blood lead concentrations, breastfeeding status, and HOME scores (measure of the quality of a child's home environment) (https://www.nlsinfo.org/content/cohorts/nlsy79-children/topical-guide/assessments/home-home-observation-measurement) were all considered in the DAG but were determined to either be intermediates or considered unnecessary based on temporal association. We also used specific gravity as a covariate in all models to adjust for urine dilution (MacPherson et al., 2018). Values for both urine metabolites and blood metals that were below the LOD were replaced with LOD/√2.

Statistical Analysis

Urinary phthalate metabolite concentrations were right skewed, so therefore they were log2 transformed. We developed linear regression models with continuous, categorical, and dichotomous variables based on the normal distribution of the log-corrected phthalate distribution. Residual plots were visually inspected to confirm assumptions of linearity and homoscedasticity. There were no excluded outliers. The primary and secondary analyses in this study were completed using complete case analysis.

To avoid biases from traditional stepwise model building, we built single-chemical models by determining potential confounders using a DAG, and selecting confounders of interest using a change in estimate (CIE) procedure (Evans et al., 2012), using the R ABE (Augmented Backwards Elimination) library (R Core Team, 2018). Study site and specific gravity were mandatory variables in all models. A 10% change in estimate and statistical significance of p<0.2 were specified for inclusion in the final model. The final model included variables for specific gravity, categorical variables for study site, mother's smoking status during pregnancy, and mother's education level. Because phthalates have been shown to impact IQ scores in a sexually dimorphic manner in other studies (Factor-Litvak et al., 2014; Kim et al., 2018), we investigated potential sex-dependent effects by including a sex and phthalate product interaction term in all single chemical models. We stratified by sex when the interaction term had a p-value <0.2.

Mixture Analysis: Weighted Quantile Sum Regression

To qualitatively assess the effect of exposure to multiple phthalates, we used Weighted Quantile Sum (WQS) regression analysis (Carrico et al., 2015). This method is used to assess associations between correlated exposures and outcomes of interest, while limiting model bias due to collinearity because of correlated exposures. WQS addresses the complexity of high dimensional data and potential collinear error in linear modeling by developing a unidirectional index based on quantiles (in our case, quartiles) of exposure. A summary term called a weighted index is developed, which is interpreted along with a set of weighted chemicals that most accurately predicts the association with the outcome, in this case child IQ. The association of each chemical to the outcome is expressed as a percentage (the weight), which can be used to determine the "bad actors" in the mixture. The index term cannot be directly interpreted in terms of magnitude of association, but gives a qualitative assessment of the general association between the chemical mixture and the outcome of interest. WQS is advantageous compared to other mixture methods because of the method's ability to simultaneously assess the mixture effect and consider confounders, while also offering a highly interpretable output.

WQS analysis was completed using the gWQS package in R Studio (R Core Team, 2018). The index was constrained in the negative direction, meaning we restricted the estimate of the association between the mixture and IQ scores to be negative. We repeated this model 100 times using the repeated holdout method (E. Tanner et al., 2019) to estimate the stability of the WQS estimate. For each repetition, 60% of the data were used for validation with 100 bootstrap samples. The full sample, with no sex stratification, was considered for the WQS analysis because of sample size restrictions. Confounders from the most parsimonious model were included in the WQS model (specific gravity, study site, maternal smoking status, and mother's education level).

To avoid concerns that the high correlation between the DEHP metabolites and weak correlations with the outcome of interest might skew results, we ran the WQS analysis considering the molar sum of the DEHP metabolites (MEHP, MEOHP, and MEHHP) as a single exposure ($\sum$DEHP). To investigate potential sex differences in the WQS models, we ran an additional model incorporating sex and sex*WQS interaction term.

Sensitivity Analysis

We performed several sensitivity analyses to investigate the reliability of the models. Additional models were tested to investigate potential confounders that were either excluded from the DAG or excluded following the CIE procedure to ensure that no major confounders that may alter the directionality of exposure were missed in the analysis (Figure S4.1). We tested models that included lead, Home Observational Measurement of the Environment (HOME) Score (Caldwell & Bradley, 2003) or breastfeeding status to determine if their inclusion would alter any observed associations (Figure S4.3). Exposure quartiles were determined for the chemicals identified in Single Chemical Models to examine the difference for the highest and lowest quartiles of exposures and for comparison to the WQS results (Table S4.3).

We also investigated a second WQS model that included all phthalate metabolites individually (MEP, MnBP, MCPP, MEHP, MEHHP, MEOHP, and MBzP).

Results
Population Characteristics

There were some differences between the full MIREC cohort and the subset identified in this study (Table 4.1). The subset population on average was more likely ($p<0.05$) to have a higher income and education level than the total MIREC population. On average, participants in this subset had significantly lower exposure to all phthalates ($p<0.05$) except MCPP as compared to the full cohort, but there were no differences in average exposure between groups for lead or OP pesticides. Phthalate metabolite levels were found to be correlated (Figure S4.5).

Single Chemical Models

In univariate and adjusted single chemical models, a doubling of mono (3-carboxypropyl) phthalate MCPP was inversely associated with FSIQ scores (β: -0.8, 95% CI: -1.5, 0.0). MCPP was also inversely associated with PIQ subtest scores (β: -1.0, 95% CI: -1.7, -0.1), but not VIQ subtest scores. There were no observed sex effects in the multivariable models when the sex interaction term was included (Table S4.1). The sum of DEHP metabolites was inversely associated with PIQ scores, although the association was not statistically significant (p<0.1). No single metabolite was significantly associated with VIQ scores in either model, and there were no positive associations between other metabolites and IQ scores that were statistically significant.

Weighted Quantile Sum

The WQS index was inversely associated with FSIQ scores (β=-1.43 95% CI: -3.37.93, - 0.02) and inversely and significantly associated with PIQ scores (β=-2.25, 95% CI: -3.70, -0.73). The FSIQ index was driven by MCPP (42%), MnBP (29%), the sum of the DEHP metabolites (14%), and MEP (12%). The PIQ index was similarly driven by MCPP (54%), MnBP (22%), and the sum of the DEHP metabolites (14%).

In the WQS models that considered sex, the mixture of phthalates was inversely associated with FSIQ and PIQ scores, but the distribution crossed zero. There was no evidence of a difference between the sex-stratified models as the distributions of all sex interaction terms also crossed zero.

Sensitivity Analysis

The inclusion of all covariates from the DAG in single-chemical models did not alter the observed associations between phthalate metabolites and IQ scores (Figure S4.2). In models that included cord blood lead exposure, breastfeeding status, and HOME score variables in the models (Figure S4.3), both

MCPP and MnBP were found to be associated with lower FSIQ and PIQ scores. However, the restricted

data set was approximately half the size of the primary models because breastfeeding status was not

collected for all participants, limiting the number of complete cases available for analysis.

For all chemicals that were associated with FSIQ, PIQ, or VIQ at a statistical significance of p<0.2 in single

chemical models (MCPP, MEOHP, and DEHP), we grouped exposure into quartiles (Table S4.3).

Individuals that were in the highest exposure level for MCPP were found to have significantly lower

average FSIQ and PIQ scores as compared to the lowest exposure group. No statistically significant

differences were observed between exposure quartiles for either DEHP or MEOHP exposure.

Discussion

The main goal of this study was to investigate the potential associations between gestational phthalate

exposure and child intelligence. I found an inverse association between first-trimester exposure to

MCPP and FSIQ and PIQ scores at age three. I also identified that MCPP, MnBP, and DEHP metabolites

may be important actors when considering the total impact of first trimester phthalate exposure and

cognitive development.

MCPP is a major metabolite of di-n-octyl phthalate (DnOP), as well as a minor metabolite of MnBP and

several long-chain phthalates (Calafat et al., 2006; Silva et al., 2007). In Canada, MCPP was detected in

over 95% of Canadians in the Canadian Health Measures Survey, a national cross-sectional

biomonitoring survey (Health Canada, 2019). The geometric mean MCPP measured in our study (0.8

µg/L) was comparable to what was observed in Canadians aged 20-39 in the CHMS cycle 5 (0.7 µg/L),

but lower than in CHMS cycle 1 (2007-2009) (1.3 µg/L) and CHMS cycle 2 (2009-2011) (1.9 µg/L).

Compared to the other metabolites measured in our study with >70% detection, the geometric mean of MCPP was lower, with over 20% of measured samples below the limit of detection (LOD).

MCPP has rarely been associated with diminished IQ scores in other studies that have considered single phthalates and their associations to IQ. In studies that have measured MCPP during the prenatal period, no significant association with IQ was observed in five studies (Li et al., 2019; Nakiwala et al., 2018; Qian et al., 2019; E. M. Tanner et al., 2019; Téllez-Rojo et al., 2013).

In contrast to the novel MCPP results in this study, DEHP is often investigated in other studies looking at phthalate exposure and cognitive development. DEHP exposure in this study was inversely associated with PIQ scores (β=-1.0, 95% CI: -2.2, 0.1) in single chemical models, although not at a statistically significantly level (p<0.1). The DEHP metabolites MEOHP and MEHHP were both inversely associated with PIQ scores, but were not statistically significant (P<0.2). We identified DEHP as a contributing chemical (14%) in the WQS models.

DEHP is a HMW phthalate that has been identified in other epidemiological studies as a risk factor for shorter male anogenital distance (Radke et al., 2018; Zarean et al., 2019) and preterm birth (Radke et al., 2019). Several studies have suggested that the metabolites of DEHP are more biologically active than DEHP (Beg & Sheikh, 2020; Kavlock et al., 2002), which is one of the reasons that both individual metabolites and the sum of metabolites were examined in this study. Despite several restrictions on DEHP usage, DEHP metabolites have consistently been detected in over 98% of participants in Cycle 1, 2, and 5 of the CHMS (Health Canada, 2019). The geometric mean of MEOHP levels measured in this study (5.5 μg/L) is lower than the levels measured for Canadians aged 20-39 in CHMS Cycle 1 (2007-2001, 13 μg/L), but comparable with the levels measured in cycle 2 (2009-2011) and 5 (2016-2017) (6.6 μg/L and 3.1 μg/L) (Health Canada, 2019). Several other studies investigating intellectual function have identified DEHP as a chemical of concern (Cho et al., 2010; Huang et al., 2015; Kim et al., 2011; Téllez-Rojo et al.,

2013), with MEOHP also specifically identified in several studies as a metabolite of concern (Cho et al., 2010; Kim et al., 2017; Kim et al., 2011; Téllez-Rojo et al., 2013).

It is unclear exactly how phthalates act in the human body to produce adverse outcomes, but toxicological evidence consistently shows that phthalates act as endocrine disruptors in animal models. Disruptions to the production and function of thyroid hormones, which are essential for neurodevelopment, have been observed in animal models (Dong et al., 2017; Ghisari & Bonefeld-Jorgensen, 2009; Sugiyama et al., 2005). Di(2-ethylhexyl) phthalate (DEHP) has been found to disrupt thyroid hormones in rats at sub-chronic levels (Poon et al., 1997) and dicyclohexyl phthalate (DCHP), n-butylbenzyl phthalate (BBzP), and di-n-butyl phthalate (DnBP) have been shown to reduce T_3 activity in vitro (Sugiyama et al., 2005). Several phthalates have also been found to act as antiandrogens (Andrade et al., 2006; Christen et al., 2012; Howdeshell et al., 2008) and may affect the sex-specific organization of neural structures.

One proposed mechanism by which phthalate exposure may alter human development is disruption of thyroid hormone (Dong et al., 2017; Ghisari & Bonefeld-Jorgensen, 2009; Sugiyama et al., 2005; Sun et al., 2018; Yao et al., 2016) Maternal thyroid dysfunction during pregnancy can cause serious impacts on brain development, and must be properly managed for optimal fetal development (Reid et al., 2010). Minor alterations in thyroid hormones may also cause changes in intellectual function (Haddow et al., 1999; Thompson et al., 2018). The human brain is sensitive to thyroid hormones during all stages of life, but because of the rapid brain development that occurs in the very early stages of development, small changes in thyroid function during the first trimester can have lasting developmental impacts (Romano et al., 2018). During human development, the fetal thyroid typically does not reach function until the second trimester (de Escobar et al., 2008; Moog et al., 2017). Prior to fetal thyroid maturation, maternal thyroid hormones are transferred across the placenta and are important for brain development

(Obregon et al., 2007). In a previous epidemiological study, Romano et al. (2018) considered exposure to a mixture of phthalates measured at 16 weeks gestation and the impacts on maternal thyroid levels and found that a mixture of phthalates at 16 weeks was significantly inversely associated with maternal TT4 levels ($\beta = -0.60$; 95% CI: -1.01, -0.18), with MEP (0.39) and MCPP (0.37) most contributing to the index. While Romano et al's (2018) study did not examine the association of phthalates with IQ scores, the identification of MEP and MCPP as metabolites of concern is consistent with the metabolites of concern identified in this study, and may suggest a biologically plausible mechanism related to brain development during the early gestational period.

The WQS method used in this study has been used in other studies that investigated prenatal or early life phthalate exposure and a variety of neurodevelopmental and hormonal outcomes(Daniel et al., 2020; Romano et al., 2018; Stroustrup et al., 2018; E. Tanner et al., 2019). Tanner et al. (2019) considered first trimester exposure to a mixture of environmental chemicals, including 15 phthalate metabolites and summed phthalate metabolites, and the overall association to IQ at age 7. Overall, the mixture was associated with a decrease in IQ for both boys and girls, and MBzP and MEP were consistently identified as chemicals of concern in the index. Daniel et al. (2020) considered the association between third trimester phthalate exposure and motor function at age 11, and found that females exposed to higher levels of non-DEHP metabolites had significantly lower fine motor scores as compared to those who had lower levels of exposure. However, this study had a small sample size and did not incorporate the repeated partition WQS method (Tanner et al., 2019).

Strengths and Limitations

As with many endocrine-disrupting chemicals, determining causality in the relationship between early-life phthalate exposure and neurodevelopment can be complicated by confounding, exposure misclassification, uncertain periods of heightened vulnerability, sexually dimorphic effects, and mixture effects (Braun, 2017). While this study aimed to account for confounding and mixture effects, the chance of residual effects cannot be discounted. Exposure classification in this study was completed using single spot samples, which is likely to have led to some exposure misclassification (Fisher et al., 2015). However, this misclassification would be non-differential, meaning it is possible that the observed associations are an underestimation.

Several demographic variables in this study were based on self-reported questionnaires, which may have been a source of reporting bias. In the MIREC study, demographic variables such as household income, smoking status, and BMI were missing for between 5-15% of participants. Our study included complete case analysis only, and these missing variables may have biased the results. Sensitivity analyses of smoking status in this cohort using plasma cotinine has suggested that overall, self-reporting of smoking status in the MIREC cohort is reliable, with some exceptions for those in higher socio-economic classes (Arbuckle et al., 2018).

One important limitation of WQS is the assumed linearity of the dose-response relationship. At least one previous study has investigated potential non-linear relationships between phthalate exposure and neurodevelopmental scores and determined that the relationship appeared to be linear (Qian et al., 2019). One toxicological study in zebrafish observed an additive toxicity relationship associated with a mixture of phthalates containing BBzP, DBP, DEHP, DIDP, DINP and DNOP compared to any of these phthalates alone (Chen et al., 2014). The same study found that for estrogenic effects, a departure from individual chemical effects was only seen when the phthalate concentrations were high. There is currently no consensus on the chemical kinetics and potential interactions in phthalate mixtures, and based on current evidence, assumptions of linearity seem to be valid for phthalate action in this context.

It is also possible that this study did not appropriately account for all critical windows of exposure. In Li et al. (2019)'s publication, prenatal MCPP exposure measured at 16 weeks and 26 weeks gestation was not significantly associated with IQ scores at ages 5 and 8, but MCPP exposure, as well as ΣDEHP, MBzP, and MEP at age 3 were all inversely associated to IQ scores. Both toxicological and epidemiological studies do suggest a plausible mechanism for phthalates' action on the developing brain during the prenatal period but it is equally feasible that there are additional periods of susceptibility to adverse effects of phthalates during brain development.

This study included testing of multiple hypotheses and did not incorporate a false discovery correction. It is important to consider that the results seen here may have been found because of the repeated testing, but the correction for multiple hypothesis testing may have resulted in type II error (Rothman, 1990)(. Based on the agreement between the single chemical models and the WQS model, we believe the results here warrant further investigation and are unlikely to be caused by chance.

Despite the limitations of this study, it also had several strengths. The MIREC study is one of the largest birth cohorts in North America with extensive questionnaire and biomonitoring data for key developmental windows. Phthalate measurements during the first trimester of pregnancy represents an important window of gestational exposure that has not been thoroughly examined in other studies. Further, the use of WQS modelling to examine the mixture effects of phthalates provides a more robust conclusion than traditional single-chemical models.

In the MIREC cohort, prenatal exposure to a mixture of phthalates may be associated with a subtle decrease in IQ scores at age 3 years. Specifically, MCPP appears to be associated with lower FSIQ and PIQ scores, and in a mixture of phthalate metabolites, MCPP, MnBP, MEP, and DEHP metabolites were all identified as potential metabolites of concern. Moving forward, it may be important to consider MCPP, a non-specific metabolite of high molecular weight phthalates, in studies investigating phthalate exposure and neurodevelopment.

Acknowledgements

The authors are grateful to the MIREC participants for their interest and participation in this study, and to the dedicated site staff for their efforts in recruitment, data collection, and data management. The MIREC Study was funded by Health Canada's Chemicals Management Plan, the Canadian Institute of Health Research (grant # MOP - 81285) and the Ontario Ministry of the Environment.

Figures and Tables

Table 4.1 – Descriptive statistics for participants in the MIREC and MIREC-CD PLUS Cohorts

Characteristic	MIREC Cohort			Current Study			p value
	N	Mean ± SD or N (%)	Range	N	Mean ± SD or N (%)	Range	
Mother's Age (years)	1983	32.2 ± 5.1	17.0-48.0	563	32.67 ± 4.62	18.0-46.0	0.05
Pre-Pregnancy BMI	1837	24.9 ± 5.6	15.1-67.0	562	25.09 ± 5.92	15.63-58.64	0.49
Mother's Education	1979			605			
High School or Less		173 (8.7)			30 (4.96)		
College or Some University		572 (28.9)			170 (28.10)		0.02
Completed Undergraduate Degree		724 (36.5)			241 (39.83)		
Graduate Degree		510 (25.7)			164 (27.11)		
Household Income (CAD)	1983			596			
Don't know, refuse to answer, no response		93 (4.7)			8 (1.34)		
<$50,000		347 (17.5)			90 (15.1)		
$50,000-80,000		398 (20.1)			145 (24.33)		0.0008
$80,001-100,000		388 (19.6)			118 (19.8)		
>$100, 000		757 (38.2)			235 (39.43)		
Smoking Status				583			
Current		80 (4.0)			15 (2.57)		
Former		463 (23.4)			152 (26.07)		0.1
Never		1072 (54.1)			383 (65.69)		
Quit during Pregnancy		115 (5.80)			33 (5.66)		
No response		253 (12.6)			N/A		
Breastfeeding Status	1983			333			
Less than 1 month		31 (1.6)			20 (6.01)		
1-6 months		90 (4.5)			53 (15.92)		0.01
More than 6 months		313 (15.8)			227 (68.17)		
Not sure or Not disclosed		91 (4.6)			33 (9.91)		
Not included in study		1458 (73.5)			N/A		
Intellectual Abilities							

							P value
Child Full Scale IQ				51.0-143.0	607	106.9 ± 13.5	51.0-143.0
Child Performance IQ				55.0-144.0	602	103.0 ±14.8	55.0-144.0
Child Verbal IQ				58.0-144.0	604	109.4 ± 13.2	58.0-144.0
Child's age at test (years)				3.0-4.1	608	3.43 ± 0.31	3.0-4.0
Environmental Exposures						Geometric Mean (GSD)	
MEP µg/L	1787	32.1 ±4.9		0.4-13000.0	563	27.3 ±4.8	0.35-6400.0	<0.(
MEOHP µg/L	1786	6.4 ±3.3		0.1-980.0	563	5.5 ±3.1	0.14-11.2	<0.(
MEHP µg/L	1773	2.3 ±3.2		0.1-340.0	559	1.9 ±3.0	0.14-4.0	0.0
MEHHP µg/L	1786	9.2 ±3.5		0.3-1200.0	563	7.8 ±3.3	0.3-17.4	<0.(
MCPP µg/L	1787	0.9 ±3.8		0.1-370.0	563	0.8 ±3.9	0.1-2.4	0.5
MBzP µg/L	1786	5.2 ±3.6		0.1 -420.0	562	4.5 ±3.3	0.1-8.8	<0.(
MnBP µg/L	1787	11.6 ±3.3		0.1 - 3100.0	563	10.5 ±3.4	0.1-22.5	<0.(
Cord lead µg/dL	1420	0.7 ± 2.4		0.003-5.2	482	0.79±1.6	0.2-3.5	0.3
DEP µg/L	1936	2.4 ± 2.6		0.7-3400.0	594	2.24±2.7	0.7-3400.0	0.1
DMP µg/L	1935	3.0 ± 3.0		0.7-190.0	594	2.95±3.0	0.71-75.0	0.1
DMTP µg/L	1933	3.1 ± 4.3		0.4-420.0	593	3.28±4.4	0.42-420.0	0.3

P values for the current study as compared to the whole MIREC Cohort

BMI – Body Mass Index, HOME Score – Home Observation Measurement of Environment Score, MBzP-mono-benzyl phthalate, MCPP - mono (3-carboxypropyl) phthalate , MEHHP- mono(2-ethyl-5-hydroxyhexyl) phthalate, MEHP- mono(2-ethylhexyl) phthalate, MEOHP-mono(2-ethyl-5-oxohexyl) phthalate, MEP-mono-ethyl phthalate, MnBP-mono-n-butyl phthalate, DEP - diethyl phthalate, DMTP – dimethylthiophosphate, DMP- dimethyl phthalate

Table 4.2: Beta estimate for Weighted Quantile Index for the association between measured phthalate metabolites and IQ Scores measured at age 3, using ∑DEHP Metabolites as a single measure

	β estimate (95% CI)	MCPP Weight	MEP Weight	MnBP Weight	∑DEHP Weight	MBzP Weight
FSIQ	-1.43 (-3.37, -0.02)	0.42	0.10	0.29	0.14	0.05
PIQ	-2.25 (-3.70, -0.73)	0.54	0.08	0.22	0.14	0.02
VIQ	0.05(-1.21, 1.30)	0.29	0.17	0.27	0.12	0.16

WQS Model adjusted for study site, mother's education level, and mother's smoking status during pregnancy. B Estimate and 95% CI estimates were determined using 100 repetitions of the WQS model. FSIQ – Full Scale Intelligence Quotient Score, PIQ-Performance Intelligence

Figure 4.1: Multivariable models investigating the association between individual log-2 transformed prenatal phthalate metabolites and IQ score at age 3 in the MIREC Cohort.

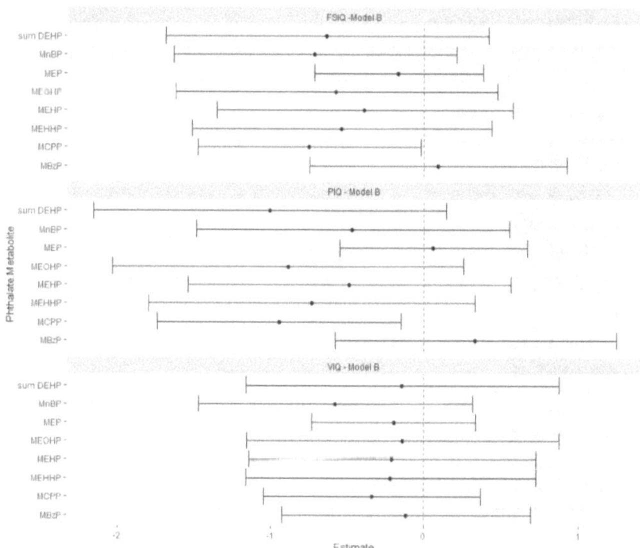

Model adjusted for specific gravity, adjusted for specific gravity, study site, maternal smoking status, and mother's education level.

MBzP-mono-benzyl phthalate, MCPP - mono (3-carboxypropyl) phthalate, MEHHP- mono (2-ethyl-5-hydroxyhexyl) phthalate, MEHP- mono (2-ethylhexyl) phthalate, MEOHP-mono (2-ethyl-5-oxohexyl) phthalate, MEP-mono-ethyl phthalate, MnBP-mono-n-butyl phthalate, DEHP – molar sum of Bis (2-ethylhexyl) phthalate metabolites

Table S4.1: Phthalate X Sex interaction terms for multivariable single-chemical models

Covariates	Metabolite x Sex Interaction Term	FSIQ β estimate (95% CI)	FSIQ p-value	PIQ β estimate (95% CI)	PIQ p-value	VIQ β estimate (95% CI)	VIQ p-value
Model B	MEP	-0.38 (-1.34, 0.57)	0.44	-0.19 (-1.26, 0.89)	0.74	-0.64 (-1.56, 0.28)	0.18
	MEOHP	0.06 (-1.28, 1.41)	0.93	-0.06 (-1.54, 1.43)	0.94	0.17 (-1.12, 1.47)	0.79
	MEHP	-0.36 (-1.74, 1.01)	0.61	-0.59 (-2.11, 0.93)	0.45	-0.09 (-1.41, 1.24)	0.89
	MEHHP	0.26 (-1.01, 1.53)	0.69	0.07 (-1.34, 1.48)	0.92	0.35 (-0.88, 1.57)	0.58
	MCPP	0.05 (-1.07, 1.16)	0.94	0.01 (-1.22, 1.24)	0.99	0.04 (-1.04, 1.12)	0.94
	MBzP	-0.33 (-1.60, 0.94)	0.61	-0.57 (-1.98, 0.84)	0.43	-0.02 (-1.24, 1.21)	0.98
	MnBP	0.60 (-0.65, 1.85)	0.34	0.31 (-1.08, 1.70)	0.66	0.66 (-0.55, 1.86)	0.29
	DEHP	0.22 (-1.82, 2.27)	0.83	0.22 (-2.04, 2.48)	0.85	0.28 (-1.69, 2.25)	0.78
Supplemental Model	MEP	-0.30 (-1.32, 0.73)	0.57	-0.32(-1.48, 0.84)	0.59	-0.47 (-1.46, 0.53)	0.36
	MEOHP	0.19 (-1.25, 1.63)	0.80	-0.28(-1.86, 1.30)	0.73	0.53(-0.87, 1.92)	0.46
	MEHP	-0.23 (-1.68, 1.22)	0.76	-0.72 (-2.32, 0.88)	0.38	0.19(-1.22, 1.60)	0.79
	MEHHP	0.36 (-0.99, 1.72)	0.60	-0.14 (-1.64, 1.35)	0.85	0.66 (-0.65, 1.98)	0.32
	MCPP	0.08 (-1.13, 1.29)	0.90	-0.13 (-1.46, 1.20)	0.85	0.21 (-0.96, 1.38)	0.73
	MBzP	-0.13 (-1.49, 1.23)	0.85	-0.59 (-2.09, 0.92)	0.44	0.30 (-1.02, 1.62)	0.66
	MnBP	0.55 (-0.77, 1.88)	0.41	-0.01(-1.48, 1.45)	0.98	0.83 (-0.46, 2.11)	0.21
	DEHP	0.34 (-1.81, 2.49)	0.76	-0.08(-2.45, 2.29)	0.95	0.70 (-1.38, 2.78)	0.51

Model a adjusted for specific gravity, study site, educational level, smoking status, cord lead levels, breastfeeding status)

Supplemental Model adjusted for specific gravity, study site, educational level, smoking status, Pre-pregnancy BMI, income level, first trimester DEP levels, First Trimester DMP level, First trimester DMTP level cord lead levels, breastfeeding status

FSIQ – Full Scale Intelligence Quotient Score, PIQ-Performance Intelligence Quotient Score VIQ – Verbal Intelligence Quotient Score

Table S4.2: Beta estimate for Weighted Quantile Index for the association between measured phthalate metabolites and IQ Scores measured at age 3

	β estimate (mean, 95% CI)	MCPP Weight	MEP Weight	MnBP Weight	MEOHP Weight	MEHHP Weight	MBzP Weight	MEHP Weight
FSIQ	-1.44 (-2.85, 0.51)	0.40	0.10	0.27	0.08	0.05	0.05	0.06
PIQ	-2.22 (-4.01, -0.39)	0.47	0.07	0.23	0.08	0.04	0.02	0.07
VIQ	0.01 (-1.07, 1.49)	0.23	0.16	0.22	0.08	0.06	0.13	0.12

WQS Model adjusted for study site, mother's education level, and mothers' smoking status during pregnancy. B Estimate and 95% CI estimates were determined using 100 repetitions of the WQS model. FSIQ – Full Scale Intelligence Quotient Score, PIQ-Performance Intelligence

Table S4.3: Results from ANOVA to investigate metabolite quartile and IQ score difference

	Quartile Comparison	IQ score difference	Simultaneous 95% Confidence Limits	Quartile Comparison	IQ score difference	Simultaneous 95% Confidence Limits	Quartile Comparison	IQ Score difference	Simultaneous 95% Confidence Limits
		MEOHP			DEHP			MCPP	
FSIQ	0 - 3	-0.03	-4.06, 4.01	0 - 3	0.04	-4.00, 4.09	0 - 3	4.62*	0.63, 8.60
	1 - 3	-1.33	-5.32, 2.65	1-3	-0.68	-4.70, 3.34	1-3	2.28	-1.70, 6.25
	2 - 3	-1.98	-5.97, 2.01	2-3	-0.95	-4.97, 3.06	2-3	2.30	-1.67, 6.27
	0 - 2	1.96	-2.06, 5.98	0 - 2	0.99	-3.04, 5.03	0 - 2	2.32	-1.66, 6.3
	1 - 2	0.65	-3.32, 4.62	1 - 2	0.27	-3.74, 4.29	1-2	-0.02	-3.99, 3.94
	0 - 1	1.31	-2.71, 5.32	0 - 1	0.72	-3.32, 4.76	0 - 1	2.34	-1.64, 6.32
PIQ	0 - 3	0.65	-3.78, 5.09	0 - 3	0.37	-4.07, 4.81	0 - 3	4.92*	0.52, 9.35
	1 - 3	-0.29	-4.67, 4.09	1 - 3	-0.09	-4.51, 4.33	1-3	3.07	-1.31, 7.44
	2 - 3	-1.79	-6.17, 2.60	2 - 3	-0.52	-4.94, 3.90	2-3	0.84	-3.54, 5.21
	0 - 2	2.44	-1.98, 6.86	0 - 2	0.89	-3.56, 5.33	0 - 2	4.08	-0.28, 8.44
	1 - 2	1.50	-2.87, 5.86	1 - 2	0.43	-3.99, 4.84	1-2	2.23	-2.10, 6.56
	0 - 1	0.94	-3.48, 5.36	0 - 1	0.46	-3.98, 4.90	0 - 1	1.85	-2.51, 6.21
VIQ	0 - 3	-0.68	-4.57, 3.20	0 - 3	-0.34	-4.23, 3.54	0 - 3	3.22	-0.62, 7.07
	1 - 3	-1.40	-5.25, 2.45	1 - 3	-0.52	-4.40, 3.35	1-3	0.75	-3.09, 4.59
	2 - 3	-1.79	-5.63, 2.05	2 - 3	-1.22	-5.07, 2.64	2-3	2.77	-1.05, 6.59
	0 - 2	1.11	-2.76, 4.97	0 - 2	0.87	-3.00, 4.74	0 - 2	0.45	-3.39, 4.29
	1 - 2	0.39	-3.44, 4.22	1 - 2	0.69	-3.17, 4.56	1-2	-2.02	5.85, 1.01
	0 - 1	0.72	-3.15, 4.59	0 - 1	0.18	-3.71, 4.07	0 - 1	2.47	-1.38, 6.33

FSIQ – Full Scale Intelligence Quotient Score, PIQ-Performance Intelligence Quotient Score VIQ – Verbal Intelligence Quotient Score

Quartiles based on urine metabolite level, adjusted for hydration levels. Lowest exposure levels are in quartile 0, and highest exposure levels are in quartile 3.

MCPP - mono(3-carboxypropyl) phthalate, MEOHP - mono(2-ethyl-5-oxohexyl) phthalate, DEHP – molar sum of Bis(2-ethylhexyl) phthalate metabolites

* P<0.05

Supplemental Figure S4.1: Directed Acyclic Graph (DAG) for potential confounders

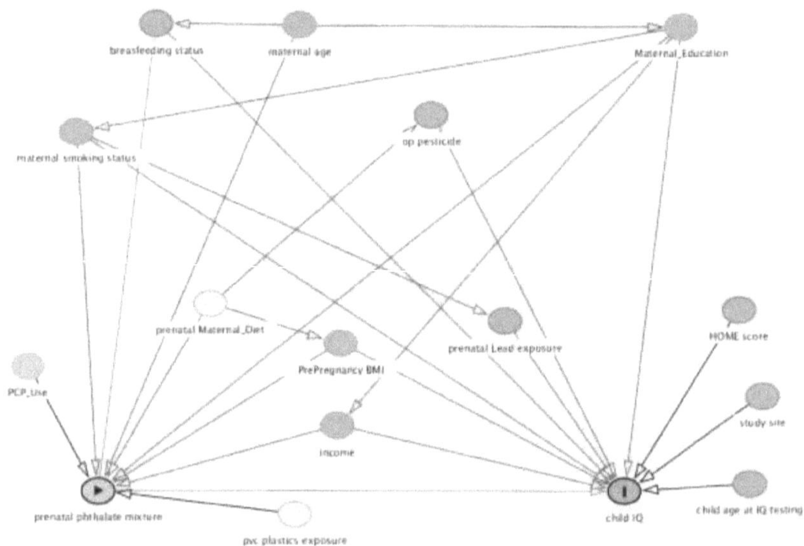

Figure S4.2: Supplemental multivariable model investigating the association between individual log-2 transformed prenatal phthalate metabolites and IQ score at age 3 in the MIREC Cohort, including all covariates and confounders identified in a DAG

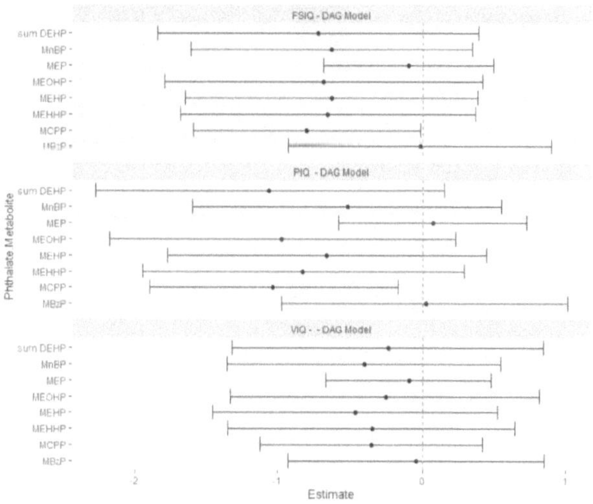

Model adjusted for Specific gravity, study site, educational level, smoking status, Pre-pregnancy BMI, income level, first trimester DEP levels, First Trimester DMP level, and First trimester DMTP level

MBzP-mono-benzyl phthalate, MCPP - mono (3-carboxypropyl) phthalate, MEHHP- mono (2-ethyl-5-hydroxyhexyl) phthalate, MEHP- mono (2-ethylhexyl) phthalate, MEOHP-mono (2-ethyl-5-oxohexyl) phthalate, MEP-mono-ethyl phthalate, MnBP-mono-n-butyl phthalate, DEHP – molar sum of Bis (2-ethylhexyl) phthalate metabolites

Figure S4.3: Supplemental multivariable models investigating the association between individual log-2 transformed prenatal phthalate metabolites and IQ score at age 3 in the MIREC Cohort, including prenatal lead exposure and breastfeeding status covariates results

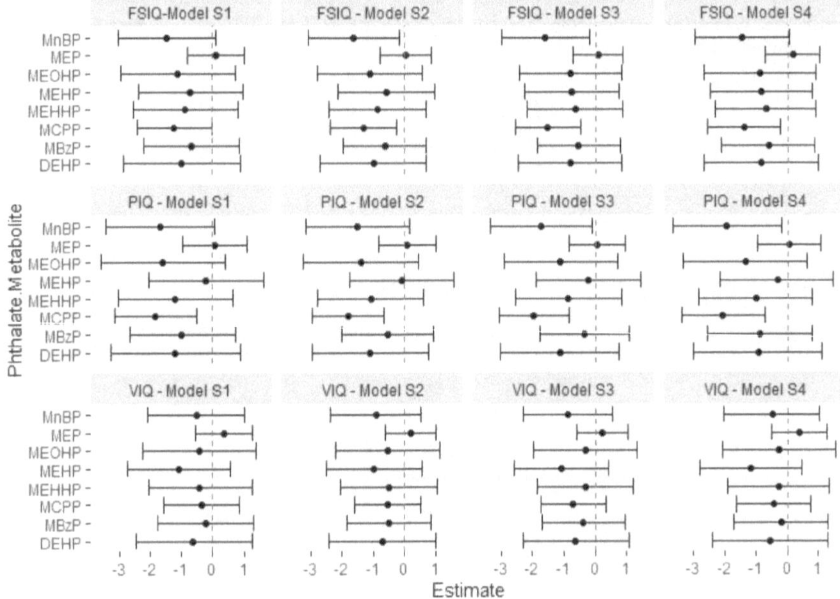

Model S1 adjusted for Specific gravity, study site, educational level, smoking status, Pre-pregnancy BMI, income level, first trimester DEP levels, First Trimester DMP level, First trimester DMTP level cord lead levels, breastfeeding status

Model S2 adjusted for Specific gravity, study site, educational level, smoking status, cord lead levels, breastfeeding status

Model S3 adjusted for HOME Score, cord lead level and breastfeeding status, specific gravity, study site, educational level, smoking status, Pre-pregnancy BMI, income level, first trimester DEP levels, First Trimester DMP level, First trimester DMTP level cord lead levels, breastfeeding status

Model S4 adjusted for HOME score, cord lead and breastfeeding status, specific gravity, study site, educational level, smoking status, cord lead levels, breastfeeding status

MBzP-mono-benzyl phthalate, MCPP - mono (3-carboxypropyl) phthalate, MEHHP- mono (2-ethyl-5-hydroxyhexyl) phthalate, MEHP- mono (2-ethylhexyl) phthalate, MEOHP-mono (2-ethyl-5-oxohexyl) phthalate, MEP-mono-ethyl phthalate, MnBP-mono-n-butyl phthalate, DEHP – molar sum of Bis (2-ethylhexyl) phthalate metabolites

Supplemental Figure S4.4: Univariate Single Phthalate Models investigating the association between log-2 transformed first-trimester phthalate metabolites and IQ score at age 3 in the MIREC Cohort.

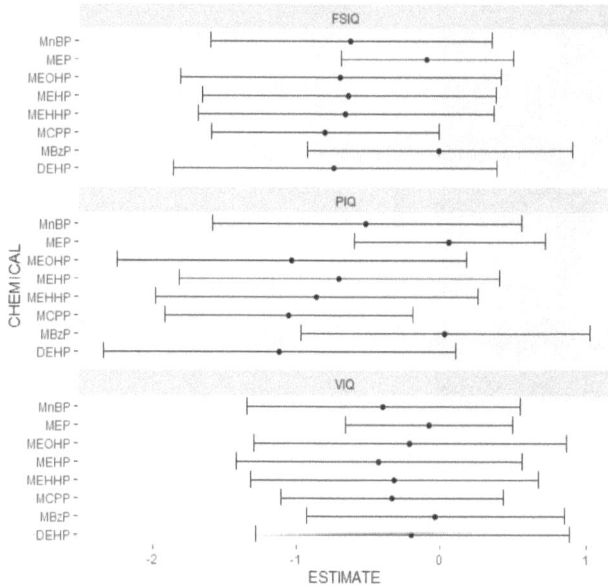

Models adjusted for specific gravity.

MBzP-mono-benzyl phthalate, MCPP - mono (3-carboxypropyl) phthalate, MEHHP- mono (2-ethyl-5-hydroxyhexyl) phthalate, MEHP- mono (2-ethylhexyl) phthalate, MEOHP mono (2 ethyl 5 oxohexyl) phthalate, MEP-mono-ethyl phthalate, MnBP-mono-n-butyl phthalate, DEHP – molar sum of Bis (2-ethylhexyl) phthalate metabolites

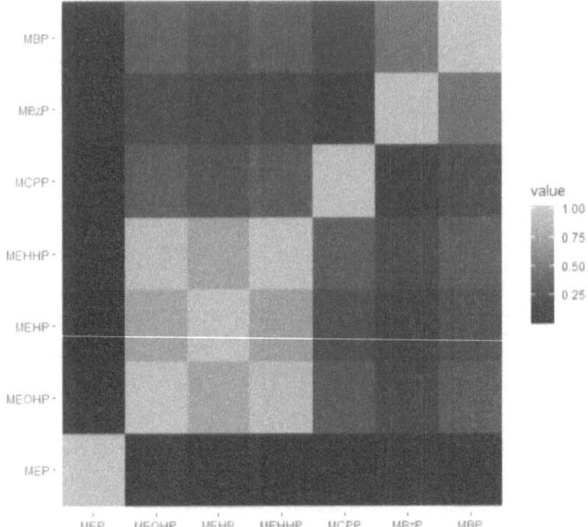
value
1.00
0.75
0.50
0.25
MBP
MBzP
MCPP
MEHHP
MEHP
MEOHP
MEP
MEP
MEOHP
MEHP
MEHHP
MCPP
MBzP
MBP

Manuscript References

Andrade, A. J. M., Grande, S. W., Talsness, C. E., Grote, K., & Chahoud, I. (2006). A dose–response study following in utero and lactational exposure to di-(2-ethylhexyl)-phthalate (DEHP): Non-monotonic dose–response and low dose effects on rat brain aromatase activity. *Toxicology, 227*(3), 185–192. https://doi.org/10.1016/J.TOX.2006.07.022

Arbuckle, T., Davis, K., Marro, L., Fisher, M., Legrand, M., LeBlanc, A., Gaudreau, E., Foster, W. G., Choeurng, V., Fraser, W. D., & Group, M. S. (2014). Phthalate and bisphenol A exposure among pregnant women in Canada—results from the MIREC study. *Environment International, 68*, 55–65. https://doi.org/10.1016/j.envint.2014.02.010 [doi]

Arbuckle, T., Fisher, M., MacPherson, S., Lang, C., Provencher, G., LeBlanc, A., Hauser, R., Feeley, M., Ayotte, P., Neisa, A., Ramsay, T., & Tawagi, G. (2016). Maternal and early life exposure to phthalates: The Plastics and Personal-care Products use in Pregnancy (P4) study. *Sci Total Environ, 551–552*, 344–356. https://doi.org/10.1016/j.scitotenv.2016.02.022

Arbuckle, T E, Fraser, W. D., Fisher, M., Davis, K., Liang, C. L., Lupien, N., Bastien, S., Velez, M. P., von Dadelszen, P., Hemmings, D. G., Wang, J., Helewa, M., Taback, S., Sermer, M., Foster, W., Ross, G., Fredette, P., Smith, G., Walker, M., … Ouellet, E. (2013). Cohort profile: the maternal-infant research on environmental chemicals research platform. *Paediatric and Perinatal Epidemiology, 27*(4), 415–425. https://doi.org/10.1111/ppe.12061 [doi]

Arbuckle, Tye E, Agarwal, A., MacPherson, S. H., Fraser, W. D., Sathyanarayana, S., Ramsay, T., Dodds, L., Muckle, G., Fisher, M., Foster, W., Walker, M., & Monnier, P. (2018). Prenatal exposure to phthalates and phenols and infant endocrine-sensitive outcomes: The MIREC study. *Environment International, 120*, 572–583. https://doi.org/10.1016/j.envint.2018.08.034

Beg, M. A., & Sheikh, I. A. (2020). Endocrine Disruption: Structural Interactions of Androgen Receptor against Di(2-ethylhexyl) Phthalate and Its Metabolites. *Toxics, 8*(4). https://doi.org/10.3390/toxics8040115

Braun, J. M. (2017). Early Life Exposure to Endocrine Disrupting Chemicals and Childhood Obesity and Neurodevelopment. *Nature Reviews. Endocrinology, 13*(3), 161–173. https://doi.org/10.1038/nrendo.2016.186

Braun, J. M., Gennings, C., Hauser, R., & Webster, T. F. (2016). What Can Epidemiological Studies Tell Us about the Impact of Chemical Mixtures on Human Health? *Environmental Health Perspectives, 124*(1), A6–A9. https://doi.org/10.1289/ehp.1510569

Calafat, A. M., Silva, M. J., Reidy, J. A., Earl Gray, L., Samandar, E., Preau, J. L., Herbert, A. R., & Needham, L. L. (2006). Mono-(3-carboxypropyl) phthalate, a metabolite of di-n-octyl phthalate. *Journal of Toxicology and Environmental Health. Part A, 69*(3–4), 215–227. https://doi.org/10.1080/15287390500227381

Caldwell, B. M., & Bradley, R. H. (2003). *Home inventory administration manual*. University of Arkansas for Medical Sciences.

Canada, H. (2013). *Second Report on Human Biomonitoring of Environmental Chemicals in*

Canada: Results of the Canadian Health Measures Survey Cycle 2 (2009-2011). Health Canada.

Carrico, C., Gennings, C., Wheeler, D. C., & Factor-Litvak, P. (2015). Characterization of Weighted Quantile Sum Regression for Highly Correlated Data in a Risk Analysis Setting. *Journal of Agricultural, Biological, and Environmental Statistics, 20*(1), 100–120. https://doi.org/10.1007/s13253-014-0180-3

Chen, X., Xu, S., Tan, T., Lee, S. T., & Cheng, S. H. (2014). *Toxicity and Estrogenic Endocrine Disrupting Activity of Phthalates and Their Mixtures.* 3156–3168. https://doi.org/10.3390/ijerph110303156

Cho, S.-C., Bhang, S.-Y., Hong, Y.-C., Shin, M.-S., Kim, B.-N., Kim, J.-W., Yoo, H.-J., Cho, I. H., & Kim, H.-W. (2010). Relationship between Environmental Phthalate Exposure and the Intelligence of School-Age Children. *Environmental Health Perspectives, 118*(7), 1027–1032. https://doi.org/10.1289/ehp.0901376

Christen, V., Crettaz, P., Oberli-Schrämmli, A., & Fent, K. (2012). Antiandrogenic activity of phthalate mixtures: Validity of concentration addition. *Toxicology and Applied Pharmacology, 259*(2), 169–176. https://doi.org/https://doi.org/10.1016/j.taap.2011.12.021

D Wechsler. (2002). *WPPSI-III Technical and Interpretive Manual.* WechThe Psychological Corporation, Harcourt Assessment Company.

Daniel, S., Balalian, A. A., Whyatt, R. M., Liu, X., Rauh, V., Herbstman, J., & Factor-Litvak, P. (2020). Perinatal phthalates exposure decreases fine-motor functions in 11-year-old girls: Results from weighted Quantile sum regression. *Environment International, 136*, 105424. https://doi.org/https://doi.org/10.1016/j.envint.2019.105424

de Escobar, G. M., Ares, S., Berbel, P., Obregón, M. J., & del Rey, F. E. (2008). The Changing Role of Maternal Thyroid Hormone in Fetal Brain Development. *Seminars in Perinatology, 32*(6), 380–386. https://doi.org/10.1053/j.semperi.2008.09.002

Doherty, B. T., Engel, S. M., Buckley, J. P., Silva, M. J., Calafat, A. M., & Wolff, M. S. (2017). Prenatal Phthalate Biomarker Concentrations and Performance on the Bayley Scales of Infant Development-II in a Population of Young Urban Children. *Environmental Research, 152*, 51–58. https://doi.org/10.1016/j.envres.2016.09.021

Dong, X., Dong, J., Zhao, Y., Guo, J., Wang, Z., Liu, M., Zhang, Y., & Na, X. (2017). Effects of long-term in vivo exposure to di-2-ethylhexylphthalate on thyroid hormones and the TSH/TSHR signaling pathways in wistar rats. *International Journal of Environmental Research and Public Health, 14*, 44. https://doi.org/10.3390/ijerph14010044

Engel, S. M., Villanger, G. D., Nethery, R. C., Thomsen, C., Sakhi, A. K., Drover, S. S. M., Hoppin, J. A., Zeiner, P., Knudsen, G. P., Reichborn-Kjennerud, T., Herring, A. H., & Aase, H. (2018). Prenatal Phthalates, Maternal Thyroid Function, and Risk of Attention-Deficit Hyperactivity Disorder in the Norwegian Mother and Child Cohort. In *Environ Health Perspect* (Vol. 126, p. 57004). https://doi.org/10.1289/ehp2358

Evans, D., Chaix, B., Lobbedez, T., Verger, C., & Flahault, A. (2012). Combining directed acyclic graphs and the change-in-estimate procedure as a novel approach to adjustment-variable selection in epidemiology. *BMC Medical Research Methodology, 12*(1), 156. https://doi.org/10.1186/1471-2288-12-156

Factor-Litvak, P., Insel, B., Calafat, A., Liu, X., Perera, F., Rauh, V., & Whyatt, R. (2014). Persistent Associations between Maternal Prenatal Exposure to Phthalates on Child IQ at Age 7 Years: e114003. *PLoS ONE, 9*(12). https://doi.org/10.1371/journal.pone.0114003

Fisher, M., Arbuckle, T. E., Mallick, R., LeBlanc, A., Hauser, R., Feeley, M., Koniecki, D., Ramsay, T., Provencher, G., Bérubé, R., & Walker, M. (2015). Bisphenol A and phthalate metabolite urinary concentrations: Daily and across pregnancy variability. *Journal of Exposure Science & Environmental Epidemiology, 25*(3), 231–239. http://10.0.4.14/jes.2014.65

Ghisari, M., & Bonefeld-Jorgensen, E. C. (2009). Effects of plasticizers and their mixtures on estrogen receptor and thyroid hormone functions. *Toxicology Letters, 189*(1), 67–77. https://doi.org/https://doi.org/10.1016/j.toxlet.2009.05.004

Greenland, S., Pearl, J., & Robins, J. M. (1999). Causal Diagrams for Epidemiologic Research. *Epidemiology, 10*(1). https://journals.lww.com/epidem/Fulltext/1999/01000/Causal_Diagrams_for_Epidemiologic_Research.8.aspx

Haddow, J. E., Palomaki, G. E., Allan, W. C., Williams, J. R., Knight, G. J., Gagnon, J., O'Heir, C. E., Mitchell, M. L., Hermos, R. J., Waisbren, S. E., Faix, J. D., & Klein, R. Z. (1999). Maternal Thyroid Deficiency during Pregnancy and Subsequent Neuropsychological Development of the Child. *New England Journal of Medicine, 341*(8), 549–555. https://doi.org/10.1056/NEJM199908193410801

Health Canada. (2019). *FIFTH REPORT ON HUMAN BIOMONITORING OF ENVIRONMENTAL CHEMICALS IN CANADA: Results of the Canadian Health Measures Survey Cycle 5 (2016–2017).* https://www.canada.ca/content/dam/hc-sc/documents/services/environmental-workplace-health/reports-publications/environmental-contaminants/fifth-report-human-biomonitoring/pub1-eng.pdf

Hernán, M. A., Hernández-Diaz, S., & Robins, J. M. (2004). A Structural Approach to Selection Bias. *Epidemiology, 15*(5). https://journals.lww.com/epidem/Fulltext/2004/09000/A_Structural_Approach_to_Selection_Bias.20.aspx

Howdeshell, K. L., Wilson, V. S., Furr, J., Lambright, C. R., Rider, C. V, Blystone, C. R., Hotchkiss, A. K., & Gray Jr, L. E. (2008). A Mixture of Five Phthalate Esters Inhibits Fetal Testicular Testosterone Production in the Sprague-Dawley Rat in a Cumulative, Dose-Additive Manner. *Toxicological Sciences, 105*(1), 153–165. https://doi.org/10.1093/toxsci/kfn077

Huang, H. B., Chen, H. Y., Su, P. H., Huang, P. C., Sun, C. W., Wang, C. J., Hsiung, C. A., & Wang, S. L. (2015). Fetal and Childhood Exposure to Phthalate Diesters and Cognitive Function in Children Up to 12 Years of Age: Taiwanese Maternal and Infant Cohort Study. In A. Gonzalez-Bulnes (Ed.), *PLoS One* (Vol. 10). https://doi.org/10.1371/journal.pone.0131910

Kavlock, R., Boekelheide, K., Chapin, R., Cunningham, M., Faustman, E., Foster, P., Golub, M., Henderson, R., Hinberg, I., Little, R., Seed, J., Shea, K., Tabacova, S., Tyl, R., Williams, P., & Zacharewski, T. (2002). NTP Center for the Evaluation of Risks to Human Reproduction: phthalates expert panel report on the reproductive and developmental toxicity of di(2-ethylhexyl) phthalate. *Reproductive Toxicology (Elmsford, N.Y.), 16*(5), 529–653. https://doi.org/10.1016/s0890-6238(02)00032-1

Kim, S., Eom, S., Kim, H. J., Lee, J. J., Choi, G., Choi, S., Kim, S. Y., Cho, G., Kim, Y. D., Suh, E., Kim, S. K., Kim, G. H., Moon, H. B., Park, J., Choi, K., & Eun, S. H. (2018). Association between maternal exposure to major phthalates, heavy metals, and persistent organic pollutants, and the neurodevelopmental performances of their children at 1 to 2years of age-CHECK cohort study. *Sci Total Environ*, *624*, 377–384. https://doi.org/10.1016/j.scitotenv.2017.12.058

Kim, Y., Ha, E. H., Kim, E. J., Park, H., Ha, M., Kim, J. H., Hong, Y. C., Chang, N., & Kim, B. N. (2011). Prenatal exposure to phthalates and infant development at 6 months: prospective Mothers and Children's Environmental Health (MOCEH) study. *Environ Health Perspect*, *119*(10), 1495–1500.

Koch, H. M., Bolt, H. M., Preuss, R., & Angerer, J. (2005). New metabolites of di(2-ethylhexyl)phthalate (DEHP) in human urine and serum after single oral doses of deuterium-labelled DEHP. *Archives of Toxicology*, *79*(7), 367–376. https://doi.org/10.1007/s00204-004-0642-4

Li, N., Papandonatos, G. D., Calafat, A. M., Yolton, K., Lanphear, B. P., Chen, A., & Braun, J. M. (2019). Identifying periods of susceptibility to the impact of phthalates on children's cognitive abilities. *Environmental Research*, *172*, 604–614. https://doi.org/https://doi.org/10.1016/j.envres.2019.03.009

MacPherson, S., Arbuckle, T. E., & Fisher, M. (2018). Adjusting urinary chemical biomarkers for hydration status during pregnancy. *Journal of Exposure Science & Environmental Epidemiology*, *28*(5), 481–493. https://doi.org/10.1038/s41370-018-0043-z

Meeker, J. D., Sathyanarayana, S., & Swan, S. H. (2009). Phthalates and other additives in plastics: human exposure and associated health outcomes. *Philosophical Transactions of the Royal Society B: Biological Sciences*, *364*(1526), 2097. http://rstb.royalsocietypublishing.org/content/364/1526/2097.abstract

Moog, N. K., Entringer, S., Heim, C., Wadhwa, P. D., Kathmann, N., & Buss, C. (2017). Influence of maternal thyroid hormones during gestation on fetal brain development. *Neuroscience*, *342*, 68–100. https://doi.org/10.1016/j.neuroscience.2015.09.070

Nakiwala, D., Peyre, H., Heude, B., Bernard, J. Y., Béranger, R., Slama, R., Philippat, C., Charles, M. A., De Agostini, M., Forhan, A., Ducimetière, P., Kaminski, M., Saurel-Cubizolles, M. J., Dargent-Molina, P., Fritel, X., Larroque, B., Lelong, N., Marchand, L., Nabet, C., ... Botton, J. (2018). In- utero exposure to phenols and phthalates and the intelligence quotient of boys at 5 years. *Environmental Health: A Global Access Science Source*, *17*(1). https://doi.org/10.1186/s12940-018-0359-0

National Research Council (US) Committee on the Health Risks of Phthalates. (2008). *Phthalates and Cumulative Risk Assessment: The Tasks Ahead.* https://doi.org/10.17226/12528

Obregon, M., Calvo, R., Escobar del Rey, F., & Morreale de Escobar, G. (2007). Ontogenesis of Thyroid Function and Interactions with Maternal Function. In *Endocrine Development* (Vol. 10, pp. 86–98). https://doi.org/10.1159/000106821

Olesen, T. S., Bleses, D., Andersen, H. R., Grandjean, P., Frederiksen, H., Trecca, F., Bilenberg, N., Kyhl, H. B., Dalsager, L., Jensen, I. K., Andersson, A. M., & Jensen, T. K. (2018). Prenatal phthalate exposure and language development in toddlers from the Odense Child Cohort. *Neurotoxicology and Teratology*. https://doi.org/10.1016/j.ntt.2017.11.004

Oulhote, Y., Lanphear, B., Braun, J. M., Webster, G. M., Arbuckle, T. E., Etzel, T., Forget-dubois, N.,

Seguin, J. R., Bouchard, M. F., Macfarlane, A., Ouellet, E., Fraser, W., & Muckle, G. (2020). *Gestational Exposures to Phthalates and Folic Acid , and Autistic Traits in Canadian*. *128*(February), 1–12.

Polanska, K., Ligocka, D., Sobala, W., & Hanke, W. (2014). Phthalate exposure and child development: The Polish Mother and Child Cohort Study. *Early Human Development, 90*(9), 477–485. https://doi.org/https://doi.org/10.1016/j.earlhumdev.2014.06.006

Poon, R., Lecavalier, P., Mueller, R., Valli, V. E., Procter, B. G., & Chu, I. (1997). Subchronic oral toxicity of di-n-octyl phthalate and di(2-ethylhexyl) phthalate in the rat. *Food and Chemical Toxicology, 35*(2), 225–239. https://doi.org/10.1016/S0278-6915(96)00064-6

Qian, X., Li, J., Xu, S., Wan, Y., Li, Y., Jiang, Y., Zhao, H., Zhou, Y., Liao, J., Liu, H., Sun, X., Liu, W., Peng, Y., Hu, C., Zhang, B., Lu, S., Cai, Z., & Xia, W. (2019). Prenatal exposure to phthalates and neurocognitive development in children at two years of age. *Environment International*. https://doi.org/10.1016/j.envint.2019.105023

R Core Team. (2018). *R: A Language and Environment for Statistical Computing*. http://www.r-project.org/

Radke, E. G., Braun, J. M., Meeker, J. D., & Cooper, G. S. (2018). Phthalate exposure and male reproductive outcomes: A systematic review of the human epidemiological evidence. *Environment International, 121*(Pt 1), 764–793. https://doi.org/10.1016/j.envint.2018.07.029

Radke, E. G., Glenn, B. S., Braun, J. M., & Cooper, G. S. (2019). Phthalate exposure and female reproductive and developmental outcomes: a systematic review of the human epidemiological evidence. *Environment International, 130*, 104580. https://doi.org/https://doi.org/10.1016/j.envint.2019.02.003

Reid, S. M., Middleton, P., Cossich, M. C., & Crowther, C. A. (2010). Interventions for clinical and subclinical hypothyroidism in pregnancy. *The Cochrane Database of Systematic Reviews, 7*, CD007752. https://doi.org/10.1002/14651858.CD007752.pub2

Romano, M. E., Eliot, M. N., Zoeller, R. T., Hoofnagle, A. N., Calafat, A. M., Karagas, M. R., Yolton, K., Chen, A., Lanphear, B. P., & Braun, J. M. (2018). Maternal urinary phthalate metabolites during pregnancy and thyroid hormone concentrations in maternal and cord sera: The HOME Study. In *International Journal of Hygiene and Environmental Health* (Vol. 221, Issue 4, pp. 623–631). Elsevier. https://doi.org/10.1016/j.ijheh.2018.03.010

Rothman, K. (1990). No adjustments are needed for multiplecomparisons. *Epidemiology, 1*(1), 43–46.

Samandar, E., Silva, M. J., Reidy, J. A., Needham, L. L., & Calafat, A. M. (2009). Temporal stability of eight phthalate metabolites and their glucuronide conjugates in human urine. *Environmental Research, 109*(5), 641–646. https://doi.org/https://doi.org/10.1016/j.envres.2009.02.004

Saravanabhavan, G., Guay, M., Langlois, É., Giroux, S., Murray, J., & Haines, D. (2013). *Canadian Health Measures Survey (2007 – 2009)*. 1–10.

Schettler, T., Skakkebæk, N., De Kretser, D., & Leffers, H. (2006). Human exposure to phthalates via consumer products. *International Journal of Andrology, 29*(1), 134–139. https://doi.org/10.1111/j.1365-2605.2005.00567.x

Serrano, S. E., Braun, J., Trasande, L., Dills, R., & Sathyanarayana, S. (2014). Phthalates and diet: A review of the food monitoring and epidemiology data. *Environmental Health: A Global Access Science Source, 13*(1). https://doi.org/10.1186/1476-069X-13-43

Silva, M. J., Samandar, E., Reidy, J. A., Hauser, R., Needham, L. L., & Calafat, A. M. (2007). Metabolite profiles of di-n-butyl phthalate in humans and rats. *Environmental Science & Technology, 41*(21), 7576–7580. https://doi.org/10.1021/es071142x

Stroustrup, A., Bragg, J. B., Andra, S. S., Curtin, P. C., Spear, E. A., Sison, D. B., Just, A. C., Arora, M., & Gennings, C. (2018). Neonatal intensive care unit phthalate exposure and preterm infant neurobehavioral performance. *PLoS ONE, 13*(3), e0193835. https://doi.org/10.1371/journal.pone.0193835

Sugiyama, S., Shimada, N., Miyoshi, H., & Yamauchi, K. (2005). Detection of Thyroid Systemâ€"Disrupting Chemicals Using in Vitro and in Vivo Screening Assays in Xenopus laevis. *Toxicological Sciences, 88*(2), 367–374. https://doi.org/10.1093/toxsci/kfi330

Sun, D., Zhou, L., Wang, S., Liu, T., Zhu, J., Jia, Y., Xu, J., Chen, H., Wang, Q., & Xu, F. (2018). Effect of Di-(2-ethylhexyl) phthalate on the hypothalamus-pituitary-thyroid axis in adolescent rat. *Endocrine Journal, 65*, 261–268. https://doi.org/10.1507/endocrj.EJ17-0272

Tanner, E., Hallerbäck, M., Wikström, S., Lindh, C., Kiviranta, H., Gennings, C., & Bornehag, C.-G. (2019). Early prenatal exposure to suspected endocrine disruptor mixtures is associated with lower IQ at age seven. In *Environment International* (p. 105185). https://doi.org/https://doi.org/10.1016/j.envint.2019.105185

Tanner, E. M., Bornehag, C.-G., & Gennings, C. (2019). Repeated holdout validation for weighted quantile sum regression. *MethodsX, 6*, 2855–2860. https://doi.org/https://doi.org/10.1016/j.mex.2019.11.008

Téllez-Rojo, M. M., Cantoral, A., Cantonwine, D. E., Schnaas, L., Peterson, K., Hu, H., & Meeker, J. D. (2013). Prenatal urinary phthalate metabolites levels and neurodevelopment in children at two and three years of age. *Sci Total Environ, 0*, 386–390.

Testa, C., Nuti, F., Hayek, J., De Felice, C., Chelli, M., Rovero, P., Latini, G., & Papini, A. M. (2012). Di-(2-ethylhexyl) phthalate and autism spectrum disorders. *ASN Neuro, 4*(4), 223–229. https://doi.org/10.1042/AN20120015

Thompson, W., Russell, G., Baragwanath, G., Matthews, J., Vaidya, B., & Thompson-Coon, J. (2018). Maternal thyroid hormone insufficiency during pregnancy and risk of neurodevelopmental disorders in offspring: A systematic review and meta-analysis. *Clinical Endocrinology, 88*(4), 575–584. https://doi.org/10.1111/cen.13550

Whyatt, R. M., Liu, X., Rauh, V. A., Calafat, A. M., Just, A. C., Hoepner, L., Diaz, D., Quinn, J., Adibi, J., Perera, F. P., & Factor-Litvak, P. (2012). Maternal prenatal urinary phthalate metabolite concentrations and child mental, psychomotor, and behavioral development at 3 years of age. *Environmental Health Perspectives, 120*(2), 290–295. https://doi.org/10.1289/ehp.1103705

Yao, H., Han, Y., Gao, H., Huang, K., Ge, X., Xu, Y., Xu, Y., Jin, Z., Sheng, J., Yan, S., Zhu, P., Hao, J., & Tao, F. (2016). Maternal phthalate exposure during the first trimester and serum thyroid hormones in pregnant women and their newborns. *Chemosphere, 157*, 42–48.

https://doi.org/https://doi.org/10.1016/j.chemosphere.2016.05.023

Zarean, M., Keikha, M., Feizi, A., Kazemitabaee, M., & Kelishadi, R. (2019). The role of exposure to phthalates in variations of anogenital distance: A systematic review and meta-analysis. *Environmental Pollution (Barking, Essex : 1987), 247,* 172–179. https://doi.org/10.1016/j.envpol.2019.01.026

Chapter 5 – Discussion and Future Directions

Research on human phthalate exposure and the impacts of that exposure on neurodevelopment is ongoing. A recent meta-analysis of 44 studies determined that there is not yet sufficient evidence to definitively link exposure to individual phthalates and neurodevelopmental outcomes (Radke et al., 2020). However, the review examined only single chemical associations and noted that the challenges of assessing correlated phthalate mixtures may be one reason for the inconsistent results that are often observed in these studies. The study presented in this book aimed to evaluate phthalate exposure using the WQS mixture analysis method, in addition to single chemical approaches, in order increase the understanding of the effects of a multi-pollutant mixture while also providing information to compare with existing literature. This newer method of analysis presents more robust findings, but is more difficult to compare to existing literature because of the differences in methods used. While most of the associations in this study were not significant, the consistent association of MCPP with decreasing PIQ scores represents a relatively novel finding, particularly considering that MCPP is not routinely examined in all studies.

This project represents an interesting data point in the growing literature related to phthalates and endocrine disruption. Work in this field is dynamic and continues to expand rapidly. The goal of this chapter is to situate the results discussed in Chapter 4 with other ongoing research, and to suggest future directions for this research.

5.1 Toxicological Evidence and Potential Human Mechanism

Toxicological evidence suggests that there may be several ways by which phthalates disrupt neurodevelopment. The proposed mechanisms of disruption have been identified in animal models, and some supporting evidence is available in human populations (Miodovnik et al., 2014). Most commonly, phthalates are believed to act on neurodevelopment via thyroid hormone disruption, but alternate theories suggest the mechanism may be related to anti-androgenic activity or glucocorticoids

interference (Braun, 2017). Research on all of these mechanisms is ongoing, and it is unclear if effects

on human intelligence are related to one single mechanism, or are impacted by several different

biochemical pathways. Identifying the specific mode of action that phthalates take in the human body is

particularly challenging because of the complex processes that are involved in the development of child

intelligence, and the composite nature of all IQ scores.

5.1.1 Thyroid Hormone Disruption

The thyroid, a gland located in the human neck, is a vital part of the endocrine system that helps to

regulate metabolic processes using Triiodothyronine (T3) and Thyroxine (T4). The thyroid is regulated by

thyrotropin/thyroid stimulating hormone (TSH). T3 and T4 are important for both normal brain

development in early life, and normal brain function through the lifespan (Bernal, 2005). The association

between major thyroid disorders during pregnancy and atypical neurodevelopment has been well

established (Momotani et al., 1984; Neale et al., 2007), and more recent evidence has determined that

small changes in T3 and T4 levels may also impact normal brain development (Henrichs et al., 2013).

The fetal thyroid is not active in the first 12 weeks of gestation and exposure to thyroid hormones

during this time occurs via placental transfer from the mother (Shepard, 1967). As such, alterations in

maternal thyroid hormone levels, both increases and decreases, have been shown to impact outcomes

for their children. In a study of children whose mothers had hypothyroidism during pregnancy, Haddow (

1999) concluded that hypothyroidism during pregnancy, even if asymptomatic, is associated with

decreased child IQ scores measured between 7 and 9 years of age. Maternal hypothyroxinemia, or low

levels of T4, measured before 18 weeks gestation has also been associated with decreased child IQ

(Ghassabian et al., 2014). In a prospective cohort study, Korevaar and colleagues (2016) found that both

low and high free T4 measured between 9 and 18 weeks gestation were associated with lower IQ in children between 5 and 9 years of age. The same study also included MRIs on a subset of participants, and observed altered brain volume, including lower levels of grey matter and cortex volume.

These established relationships between thyroid hormones and IQ are important to consider because several studies have shown that phthalates interact with thyroid hormones in both human and animal models; this interaction is a plausible mechanism of action for the association between phthalates and child IQ. Studies in pregnant populations concluded that exposure to individual phthalates is associated with both hypothyroidism and hypothyroxinemia (Gao et al., 2017; Huang et al., 2016; Yao et al., 2016) Mixtures of phthalates have been shown be associated with decreased maternal T4, decreased TSH in newborns, and decreased cord serum T4 (Romano et al., 2018). At least one study has also looked longitudinally at prenatal (second trimester) and early childhood (age 2, 4, 6) phthalate exposure and concluded that phthalate exposure is associated with altered thyroid activity (Kim et al., 2020). In that longitudinal study, MnBP exposure during pregnancy was associated with lower TSH and free T4 levels in 6-year-old children.

Figure 5.1 – Potential phthalate-mediated disruptions in the hypothalamus-pituitary-thyroid axis

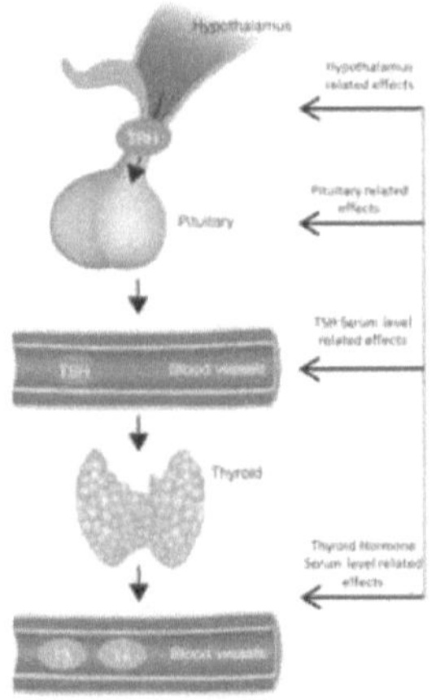

Figure modified from Oliveira et al., 2019

Evidence from animal models and in vivo analysis supports the theory that phthalates may disrupt thyroid hormone activity via a variety of pathways through the entire HPA axis (See Figure 5.1 above modified from Oliveira et al., 2019) . BBzP, DBP, and DEHP have all been shown to block T3 activity in cell models (Ghisari & Bonefeld-Jorgensen, 2009), and block the activity of a crucial gene for thyroid hormone synthesis (TRβ gene) (Sugiyama et al., 2005). More recent work has suggested that phthalates play a role in disrupting thyroid action via the TSH/TSHR (thyroid signalling hormone/thyroid signalling hormone receptor) signalling pathway (X. Dong et al., 2017; Sun et al., 2018). Any disruption in this tightly controlled biochemical pathway may lead to both downstream and upstream effects, providing biochemical plausibility for the epidemiological phenomenon seen in this study.

5.1.2 Anti-Androgenic Activity

Phthalates are known to interfere with testosterone production in Leydig cells in animal models (Foster, 2005). This anti-androgenic activity is typically associated with effects on the male reproductive system (Hannas et al., 2011), and has also been hypothesized to be associated with alterations in neurodevelopment (Weiss, 2012). Androgens in mouse models are important for sex-related brain development (Colciago et al., 2006).

In epidemiological studies, phthalate exposure during pregnancy has been linked to sex-based differences in play behaviour (Swan et al., 2010), externalizing behaviour (Engel et al., 2010) and IQ scores (Cho et al., 2010). The hypothesis is that exposure to phthalates during pregnancy would limit testosterone production by the fetus, which would then limit the substrate that is required for masculinization in the brain (Weiss, 2012). However, it is unclear if this process would affect IQ, and the majority of the research that has been completed on anti-androgenic activity in the brain is in animal models.

5.1.3 Glucocorticoids interference

Glucocorticoids are a group of steroid hormones that include cortisone, and are important in a variety of functions in the body, including the growth and normal development of the fetal brain (Moisiadis & Matthews, 2014). In the brain, glucocorticoids can inhibit normal neural differentiation during gestation (Antonow-Schlorke et al., 2003; Noorlander et al., 2014), and are also hypothesized to mediate fetal neurodevelopment via epigenetic changes (Wood, 2013).

Glucocorticoids are most often associated with maternal stress, but epidemiological evidence has suggested that phthalates also may alter the typical production and metabolism of glucocorticoids (Araki et al., 2017; Jensen et al., 2015; Sun et al., 2018). There are disagreements in exactly how phthalates interfere with normal glucocorticoid activity; while one prospective cohort study found MEHP is inversely associated with cord blood cortisone and cortisol (Araki et al., 2017), another case-control study found that there was a positive association between MECPP and cortisol levels in amniotic fluid (Jensen et al., 2015). Both MEHP and MECPP are metabolites of DEHP, and it is unclear why the results of these studies do not agree. In animal models, there is evidence that agrees with the inverse association between MEHP and glucocorticoids. In mice, MEHP inhibits a specific enzyme (11β-hydroxysteroid dehydrogenase 2), which may lower intracellular levels of certain glucocorticoids (Hong et al., 2009; Zhao et al., 2018).

5.2 Beyond Phthalates - Other chemicals of interest

This study aimed to consider some secondary chemical exposures that may have an impact on human neurodevelopment, including OP Pesticides and lead. However, in the MIREC cohort, there are several chemicals that are weakly correlated with phthalate exposure (W.-C. Lee et al., 2017) that were not included in this study and may be important to consider in future studies. It is also important to note that only the phthalate exposures were considered in the WQS analysis, but it is possible that several of these EDCs may act in a biologically additive manner, and it may be important to look at multiple chemical exposures in future studies.

5.2.1 Bisphenol-A and Analogues

Bisphenol-A (2,2-bis(4-hydroxyphenyl)propane, BPA) and replacement chemicals Bisphenol-F (BPF) and Bisphenol-S (BPS) are common plastic additives that are often correlated with phthalate exposure (Philips et al., 2018). In the MIREC Cohort, a complete-case analysis of participants with data on environmental chemicals (n=1332) determined that BPA was strongly positively correlated with MCPP (Pearson correlation coefficient 0.74), and moderately negatively correlated with MEP (Pearson correlation coefficient -0.5). BPA has been linked to neurodevelopment concerns in early childhood (Braun, 2017) and may be an important chemical to consider in future studies on EDC mixtures. The newer BPA analogues (BPF, BPS) may be of particular concern, as data on the safety of these compounds is less robust than the historical data for BPA. One recent study investigating exposure to chemical mixtures and associated childhood IQ determined that from a mixture of 26 chemicals, BPF was one of the chemicals that most contributed to a decrease in IQ scores using WQS models (Tanner et al., 2019)

5.2.2 Polybrominated Diphenyl Ethers (PBDEs)

PBDEs, a class of chemicals that are typically used in flame retardants, have been consistently linked to neurodevelopment (Lam et al., 2017); although they were not found to be strongly correlated with phthalates in the MIREC study (Lee et al., 2017). PBDEs and phthalates are both found in a variety of household goods. In the HOME study, there is some evidence of weak to moderate positive correlation between the two chemical classes (Woods et al., 2017). An economic review in the US estimated that exposure to PBDE was related to 11 million lost IQ points and over 40,000 cases of intellectual disabilities, at an excess cost of over 266 billion US dollars (Attina et al., 2016). In future studies that take a broader approach to all chemical exposures, PBDEs will be an important chemical group to consider.

5.2.3 OP Pesticides

Prenatal and early childhood exposures to OP pesticides have been consistently associated with lower scores on IQ tests (Bouchard et al., 2011; Grandjean & Landrigan, 2014; Jurewicz & Hanke, 2011). Although OP pesticides were included in some of the individual chemical models, they were not included in the WQS models in this study. In single chemical models in this study, OP pesticide exposure was not significantly associated with IQ scores, and a previous MIREC study investigating the associations between OP pesticides and IQ scores only found inverse associations among verbal IQ scores in boys (Ntantu Nkinsa et al., 2020). Exposure to OP pesticides and phthalates have both been associated with diet which may lead to correlated exposure (Pacyga et al., 2019; Papadopoulou et al., 2019; Schecter et al., 2013; van den Dries et al., 2018); but in the MIREC study, only weak correlations were observed (Lee et al., 2017). OP pesticide exposure in the United States has also been associated with profound excess loss of IQ points; estimated in the US to be associated with 1.8 million lost IQ points and over 7500 cases of intellectual disability (Attina et al., 2016).

5.2.4 Lead

Lead was included in supplementary analysis in this project, but not in the primary analysis. Lead is a known neurotoxicant during both the prenatal and childhood period (Lanphear et al., 2002) In the MIREC cohort, participating mothers with higher estimated intakes of dietary calcium, vitamin D, and iron were found to have lower blood lead levels measured in the third trimester and in cord blood (Arbuckle et al., 2016); older mothers in the MIREC study were also more likely to have higher blood lead levels (Arbuckle et al., 2016), and higher levels of MBzP and MEHHP (Arbuckle et al., 2014). MIREC mothers born outside of Canada were also more likely to have higher blood lead levels (Arbuckle et al 2016) and higher levels of MEHHP, MEHP, MEOHP and MEP (Arbuckle et al, 2014).

Prenatal lead exposure in the MIREC cohort was low, but has been found to be associated with lower PIQ scores in boys (Desrochers-Couture et al., 2018). There is consensus that there is no safe level of lead exposure for young children (EFSA Panel on Contaminants in the Food Chain, 2010). Although there have been limited associations found between lead exposure and IQ in the MIREC cohort, lead was included in the supplemental analysis that investigated lead and breastfeeding exposures. The inclusion of these two variables limited the sample size of the investigation, but did not show any notable differences compared to the main analysis.

5.3 Exposure Timing

This study specifically considered environmental exposures that occurred during the first trimester of gestation. Early pregnancy is an important period of fetal development, but it there is also evidence that other periods of development including pre-conception (J. Braun et al., 2017; Toivonen et al., 2017; Vohr et al., 2017) late pregnancy (Factor Litvak et al, 2014; Tellez-Rojo et al, 2013; Kim et al, 2011; Doherty, 2017; Whyatt et al, 2012) and early childhood (Kim, 2017; Huang et al, 2015; Cho et al, 2010) impact neurodevelopmental outcomes. In other studies, phthalate exposure has been found to be inversely associated with a variety of neurodevelopmental outcomes when exposure is measured in the third trimester (Factor Litvak et al, 2014; Tellez-Rojo et al, 2013; Kim et al, 2011; Doherty et al, 2017; Whyatt et al, 2012), during infancy (Kim, 2018), and during early childhood (Kim, 2017; Huang et al, 2015; Cho et al, 2010).

In the first 6 months of life, phthalates in breast milk and formula may be a major route of exposure for infants. In the P4 study of pregnant women from Ottawa, Canada, phthalate metabolites MEHP, MnBP, MEP, and MMP were found in all of the collected breast milk samples (Arbuckle et al., 2016) and parent compounds have also been identified in breast milk samples collected as part of the MIREC study (Cao et al., 2021). Phthalate metabolites in breast milk have all readily been identified in other study populations across the world (Fromme et al., 2011; S. Kim et al., 2018; Schlumpf et al., 2010). There have been a limited number of studies examining how breastfeeding exposure to phthalates has impacted neurodevelopment, although one small study in a Chinese population concluded that childhood exposures to phthalates were associated with increased delay in neurodevelopment at 9 months, but that breastfeeding or bottle feeding was not a major source of phthalate exposure (R. Dong et al., 2019). However, a study in a Korean cohort identified DEHP in breast milk as inversely associated with MDI scores, an analogue of IQ scores in older populations, in children at age 13 to 24 months.

The study presented in this book did not take into account later childhood phthalate exposures, which have been identified as inversely associated with IQ scores in previous studies (Cho et al., 2010; Huang et al., 2015; J. I. Kim et al., 2017; Li et al., 2019). A recent study in an American population examined phthalate exposure at eight time points during early life (twice during gestation and six times between ages 1 and 8) and the association with child IQ scores (Li et al., 2019). In that study, urinary concentration of ΣDEHP, MBzP, MEP, and MCPP metabolites measured between age 2 and 3 years was inversely associated with IQ scores, whereas MBzP measured at 16 week gestation was the only gestational exposure in that study to be significantly inversely associated with IQ. This study found no association with first trimester MBzP exposure and IQ scores, but identified three of the same metabolites of concern (MCPP, ΣDEHP, and MEP) as associated with a lower IQ score when considered using WQS. It is unclear if the identification of similar sets of chemicals of concern during different time periods in different population samples is due to spurious association, or may be an interesting point for future investigations. Urine samples were collected from 200 MIREC child participants between ages 2 and 3, and analyzed for phthalate metabolites, and these samples may be able to provide some additional information on important windows of consideration for phthalate exposure in future analyses.

Pinpointing key windows of exposure during gestation is challenging in the current literature. If single

spot samples are used, the data are likely prone to misclassification (Perrier et al., 2016); but if multiple

samples across trimesters are used, then it is impossible to determine the importance of exposure

timing. Current studies that investigate IQ or BSID scores (an IQ analogue for younger children) vary in

exposure timing and there is no obvious trend in the resulting associations. Of those studies that use

multiple samples across trimesters, the results have typically been null associations. For example, in the

CHAMACOS cohort, the relationship between phthalates and IQ scores at ages 7 to 10 was investigated

(Hyland et al., 2019). Phthalates were measured in both the first and second trimester and were

averaged as a means of estimating total exposure throughout pregnancy; however, this strategy

prevents us from determining if a specific window of exposure is important. Most of the associations

found in this study were null, although the doubling of high molecular weight phthalates ΣHMW (MBzP,

MCPP, MCOP, and MCNP) was associated with lower working memory in boys ($\beta=-2.1$, 95% CI: -4.2,

0.0) but not in the full sample or in girls. For future studies, a tool such as the Biomarker Reliability

Assessment Tool (BRAT) (Verner et al., 2020) may allow researchers to more appropriately assess how

many urine samples may be required to appropriately classify an individual's exposure to non-persistent

chemicals.

5.4 Analysis and Statistical Techniques

5.4.1 New statistical techniques

The field of mixture analysis is still evolving and development of statistical methods is an ongoing

activity. The WQS method used in this project is still relatively new for use in high dimensional exposure

studies, and is believed to sufficiently handle collinear exposures. However, since this project began,

there have already been several proposed extensions to the method to reduce bias and improve

prediction, and at least one method has been developed that may outperform WQS.

Quantile g-computing, proposed by Keil and colleagues (Keil et al., 2020) was developed to overcome some of the limitations of WQS, including the required assumption of homogenous directionality of exposure effect. The quantile g-computing method is also able to assess non-linear associations of individual components of a mixture, which is not possible using WQS. In simulation studies, Quantile g-computing limits bias as compared to WQS by handling departures from homogeneity and providing a more realistic estimate of the effect of a whole mixture. Quantile g-computing methods have not yet been widely used in environmental epidemiology studies, but may provide a viable next step for continued investigations in to the population health risks of chemical mixtures.

5.4.2 Model Comparisons

Assessing the health impacts of environmental impacts remains complicated because of the variations in modelling techniques. Direct model comparison when different techniques are used is also complicated. For example, the recent study investigating thyroid hormone disruption and phthalate exposure in a Korean population identified MnBP as a potential metabolite of concern using a BKMR model (K.-N. Kim et al., 2020); but it is unclear if the same chemicals would have been identified using a linear modeling technique such as WQS. In the literature, there have been at least two studies that used different statistical techniques to assess the associations between chemical mixtures and an outcome of interest (Chiu et al., 2018; Y. Zhang et al., 2019). These studies found that selecting different statistical tools may lead to some differences in conclusion; but both studies concluded that no one method was preferred over another and each had their own strengths and weaknesses, a conclusion that has also been echoed elsewhere (Joseph M Braun et al., 2016).

While these comparisons typically identify similar chemicals of concern, it is still possible that the methods may result in inappropriate chemical identification, or incorrect magnitudes of association. In our case, WQS was useful to identify chemicals of concern from a risk management perspective, but the method is not helpful in terms of quantifying the specific risk of IQ loss associated with exposure to a phthalate mixture.

5.4.3 Model Building Techniques

Statisticians have advocated for limiting the use of traditional stepwise model building techniques (Greenland, 1989; Mickey & Greenland, 1989), and yet model building using stepwise regression and p-value cut offs for variable inclusion still dominates many of the published papers in the field of environmental epidemiology. Greenland has argued that if the goal of model-building is to produce effect estimates that are as valid and precise as possible, then epidemiologists must ensure that they are using the best modelling strategies to avoid bias (Sander Greenland et al., 2016).

This study combined DAGs for designing the models tested (Sander Greenland et al., 1999), as well as change-in-estimate (CIE) model building techniques. Though CIE has been criticised for failing to maximize model accuracy, it is also a method that allows the model-builder to determine the criteria that results in covariate exclusion (Greenland & Pearce, 2015) . In addition, using both a DAG and CIE technique has been advocated to reduce bias while still considering extra-statistical information (Evans et al., 2012; Weng et al., 2009). Evans and colleagues (2012) argued that the combination of both DAGs and CIE procedure is advantageous because it forces explicit discussions of assumptions using a DAG, while also leading to a parsimonious model because of the incorporation of the CIE procedure.

In an attempt to avoid over adjustment bias (Schisterman et al., 2009), I focussed the model building technique on the minimal set of covariates for the WQS analysis. However, in addition to the main models presented in Chapter 4, additional models were tested to investigate potential confounders that were either excluded from the DAG or excluded following the CIE procedure to ensure that no major confounders that may alter the directionality of exposure were missed in the analysis. These models investigated the influence of potential confounders such as breastfeeding status and HOME score, both of which were measured after the initial phthalate exposure, and prenatal lead exposure, which was found to be associated with PIQ in boys in a previous MIREC study (Deroches-Couture, 2018). As seen in the single-chemical analysis with the minimal set of covariates, an increase in MCPP was also found to be associated with decreases in FSIQ and PIQ scores in the supplementary models that included breastfeeding status and HOME score as covariates. MnBP was also identified in the supplementary models as inversely associated with IQ scores. Because the subset used for the supplementary models was approximately half the size of the main models, it is unclear if this association is spurious or may be important for consideration in future studies. MnBP was also identified in the WQS model as a potential metabolite of concern, and in a previous study, MnBP exposure during pregnancy has been linked to disruptions in thyroid hormone regulation (K.-N. Kim et al., 2020).

5.5 Chemicals of Interest

5.5.1 MCPP

MCPP was found to be the only metabolite consistently associated with a decrease in FSIQ scores and PIQ scores in all models. MCPP is a non-specific by-product of several of the long-chain phthalate metabolites, and because of this non-specificity, it has not been considered in most previous studies that look at prenatal exposure and IQ scores. MCPP is a major metabolite of DnBP and DnOP, as well as a secondary metabolite of several other long-chain phthalates (e.g., DiNP, DiDP). For this reason, it is difficult to know exactly which routes of exposure would contribute MCPP exposure, but DnOP is typically found as a replacement for DEHP in PVC plastics , building materials, and food packaging (Fierens et al., 2012; Schecter et al., 2013).

In US populations, there has been an increasing trend in MCPP exposure in adults between 2001 and 2010 (A. Zota et al., 2014). In the Canadian population, children (ages 3 to 11 in cycle 2 and 5, ages 6 to

11 in cycle 1) in the CHMS were found to have significantly higher MCPP exposure as compared to other age groups (Health Canada, 2019; Saravanabhavan et al., 2013).

MCPP exposure has been associated with several adverse birth outcomes in epidemiological studies, such as decreased length of gestation (Chin et al., 2019), and higher odds of preterm birth (Ferguson et al., 2014). Increasing MCPP exposure has also been associated with decreased free T3 hormone during pregnancy (Lauren E Johns et al., 2015). However, MCPP has not been included in analysis in several other studies that examined phthalate exposure and thyroid hormones (Kim et al., 2020), making it difficult to determine the validity of this single finding and potential mechanisms of action for phthalates.

5.6 Study Strengths and Limitations

This book project has many strengths, including a large data set in the MIREC cohort, advanced model-building techniques, and relatively novel mixture analysis techniques. Despite these strengths, there are some limitations that must be considered. Specifically, biases associated with observational research, exposure misclassification, and self-selection are important factors when considering the external and internal validity of the findings in this study.

Exposure to phthalates during the first trimester was estimated using single spot urine samples. This method of phthalate exposure assessment has been criticized for being unreliable because of the short half-life of phthalate metabolites in the body (Johns et al., 2015). While some studies completed using pregnancy cohorts have found moderate to high correlation in spot samples taken within 24 hours of each other (Fisher et al., 2015), others have found that spot samples have poor to moderate correlation when samples were taken 4 weeks apart (Cantonwine et al., 2014). The P4 study (Fisher et al., 2015) investigated phthalate metabolite concentrations and temporal variability in a Canadian pregnant population with comparable age, household income, and inclusion criteria to the whole MIREC study. In the P4 study, urine MEP and MBzP metabolite levels were found to have moderate reproducibility between samples taken in the same day, and across each trimester. Both of these chemicals are short chain phthalate metabolites, and typically associated with personal care product use. On the other hand, medium and long chain phthalate metabolites typically associated with food packaging materials (MCPP, DEHP metabolites) were found to have poor reproducibility within day and across trimesters. Studies that have looked at spot samples across trimesters have found low to moderate reproducibility for all phthalate concentration across trimesters (Qian et al., 2019). MCPP was found to have a significant effect on IQ in our study; however, the reproducibility of MCPP measurements in different spot samples was found to be low in the P4 study in both within-day (ICC= 0.21 and 0.31) and across pregnancy (ICC=0.19) samples (Fisher et al., 2015), but moderate in across-day samples in the Columbia Center for Children's Environmental Health (CCCEH) cohort (ICC=0.44) (Adibi et al., 2008) and the Fox River Environment and Diet Study (FRIENDS) (ICC=0.59) (Peck et al., 2010). This study looked at both short and long-chain phthalates, in only one trimester of pregnancy. While it is possible that there was some misclassification due to the spot samples, we feel confident that this misclassification is likely non-differential, and the effect reported is likely to be biased towards the null.

This project used WPPSI-III IQ scores as the outcome of interest to examine the relationship between environmental exposures and a functional measure of child neuropsychological development (Forns et al., 2012). IQ scores are not an exclusive measurement of brain development, and are known to be associated with both demographic factors and brain development (Gale et al., 2006; Lange et al., 2010; Reiss et al., 1996; van den Dries et al., 2018). IQ scores as a measure of cognitive function in children have been criticised because of concerns about cultural bias (Braaten & Norman, 2006; Shuttleworth-Edwards, 2016). However, the WPSSI-III also has several advantages over other intelligence tests, particularly for younger Canadian preschoolers (Gordon, 2004). WPPSI-III specifically includes a test series for children between ages 2 years, 6 months, and 3 years, 11 months, in order to deal with the differences in appropriate tests for younger and older preschoolers. The WPPSI-III also specifically includes Canadian norms to deal with potential cultural and social differences as compared to a standard American samples, and was designed to be shorter and more engaging for younger pre-schoolers. Finally, the WPPSI demonstrates excelled test-retest reliability, as well as inter-rater reliability between assessors (0.98-0.99 in the standard sample of 1700) (Wechsler, 2002).

The MIREC Study measured 11 metabolites, which does offer a good picture of general exposure in the study population. However, other known metabolites exist for the parent compounds of interest for which methods of measurement have recently been developed, or for which no methods to measure yet exist. This is a major limitation in being able to extrapolate to which parent compounds may be priority for restriction or regulation. In addition, exposure measured in MIREC may not represent the phthalates to which pregnant Canadians are exposed today, because some new chemicals (e.g., DiNCH, DEHTP) have been developed as replacements for DEHP phthalates.

The MIREC study represents one of the largest birth cohorts in North America and has extensive questionnaire and biomonitoring data from key developmental windows. However, there are some limitations to this study that may be important to consider when looking at the external validity of the results. The MIREC study participants were not nationally representative, and were more likely to be Caucasian and more likely to be born in Canada than the Canadian obstetric population in general (Arbuckle et al, 2013). Additionally, MIREC participants were often urban residents because of the requirement to visit a major hospital for care and study follow-up, which may mean that this study may not appropriately capture exposure and outcomes of rural Canadian populations.

When considering external validity, in US populations, cross-sectional studies suggest that immigrant populations have higher phthalate exposure (Mitro et al., 2015), which may mean that the exposure levels seen in this study are not reflective for all population groups. In the MIREC cohort, MCPP metabolite concentrations did not significantly differ between mothers born inside and outside of Canada, but measured levels of MEP, MBzP, MEHP, MEOHP, and MEHHP metabolites were all higher in mothers born outside of Canada than those inside of Canada.

Every effort was made in this study to limit potential bias; however, as a subset of an observational pregnancy cohort, several potential weaknesses must be noted. This study utilized complete cases and a subset of the entire MIREC population, and as such, may have some element of inclusion bias. However, several different data sets of varying sample sizes were used to examine different models, and the same chemicals of interest were identified in all models.

Despite all of the limitations listed here, every attempt was made to control bias using the tools available for observational studies. Modelling decisions were based on thorough research and well-designed DAGs. The robust data collected in the MIREC study allowed for adjustment of all major confounders. It is the author's opinion that despite the limitations of this study, it still provides valuable insight in to the potential associations between prenatal exposure to some phthalates and young child intelligence.

5.7 Moving Forward: Economic Perspectives and Environmental Chemical Regulation

A recent cost-analysis was performed to estimate the costs and disease burden associated with all environmental chemical exposure and IQ loss (Gaylord et al., 2020). This study estimated the economic cost of prenatal exposure to PBDEs, OP pesticides, methylmercury, and early life exposure to lead (up to age five) between 2001 and 2016 in the United States, and found that IQ loss associated with exposure to these four chemicals resulted in an economic cost of over 6 trillion dollars.

Typically, intellectual disability is defined as an IQ score of less than 70. While the loss of approximately 1-2 IQ points estimated in this analysis may seem minor in isolation, when considering this difference as a shift in the entire normal distribution of scores, it represents a significant economic loss from a societal perspective. Previous economic assessments that investigate the societal cost of a loss of IQ have estimated the value of each lost IQ point to be 22,268 dollars and each new case of intellectual disability as a lifetime societal economic loss of 1,272,470 dollars (Gaylord et al, 2020)

While the study presented in this book overall suggests limited contributions from phthalates to neurodevelopmental outcomes, there is growing evidence that phthalates have an impact on public health, including obesity, diabetes, endometriosis, and even cardiac death (Attina et al., 2016); in the US alone, phthalates exposure is believed to have resulted is hundreds of billions of dollars of attributable costs related to diabetes, obesity, and early mortality (Attina et al., 2016). While the evidence for phthalate association to neurodevelopment is still limited, it is important to continue work in this field. Estimates from 2010 in the USA suggest that diseases related to EDCs may account for costs of close to 2% of the annual GDP (Attina et al., 2016) and 1.28% of annual GDP in the EU, representing a cost of 17.0 trillion Euros (Trasande et al., 2015).

Based on phthalates' known detrimental effects on child health, some regulations exist to limit the exposure of high-risk groups to phthalates. In the United States, the use of DEHP, DBP and BBzP in children's toys and childcare articles was banned in the Consumer Product Safety Act (*Consumer Product Safety Improvement Act of 2008*, 2008). The ban was expanded in 2017 to include DiNP, DIBP, DCHP, di-n-pentyl phthalate (DPENP), di-n-hexyl phthalate (DHEXP) (Consumer Product Safety Commission, 2017). In the EU, the use of DINP, DEHP, DBP, DIDP, DNOP, and BBzP in children's articles was banned on an emergency basis starting in 1999 (European Commission, 1999), with formal limitations in 2004 (European Parliament, 2005).

In Canada, DEHP is on the List of Toxic Substances (Canadian Environmental Protection Act , 1999) is limited or banned from being used in a variety of products, and there are limitations of the amounts of

DEHP, DBP, BBzP, DINP, DIDP and DNOP allowed in children's toys and childcare products (Canada Consumer Product Safety Act Phthalate Regulations, 2016). Canada does not have any official exposure limits for phthalate exposure during pregnancy. This may be an area for ongoing work or regulation.

Moving forward in a post-COVID-19 world, it is particularly important to continue to monitor emerging issues in phthalate exposure, for both neurodevelopment and other health effects associated with EDCs.

5.8 Conclusion

There is still no consensus on the safety of phthalates with respect to neurodevelopment. A recent review (Radke et al., 2020) highlighted the inconsistencies in the current literature, concluding that there was some moderate evidence that BBzP was associated with motor ability score, but only slight or inconsistent evidence that DEHP, DINP, DBP, DIBP, DEP, or DEP were associated with any neurocognitive score. This review did not examine MCPP. Part of the challenge of studying phthalates and neurodevelopment is the diverse group of outcomes that can be examined, and the ever-growing number of chemicals and metabolites included in the group. Human neurodevelopment and neurocognitive functioning continue to be areas with large amounts of unanswered questions, and more work is still to be done to understand what role, if any, phthalates play in the optimal functioning of the human brain.

This book contributes to the current scientific literature by examining both individual and cumulative phthalate exposure using newer statistical techniques. The results from this analysis suggest that gestational exposure to MCPP, a non-specific metabolite of long-chain phthalates, may be inversely associated with childhood IQ scores, as measured using the WPPSI-III. Sex of the child was not found to be a significant factor in this study. This result was consistent in both the individual models and the cumulative models, and in all subsets of the data that were examined. MCPP is not a major metabolite specific to one single parent phthalate, and as such, has often been left out of other analyses.

Environmental mixture analysis is a rapidly evolving field that seeks to answer many complicated

questions about the health effects of the complex exposures that humans encounter daily. While 2020

has been dominated by discussions on infectious disease epidemiology, it is important to continue to

push forward in the fields of chronic exposures to maintain population health and safety for some of the

most vulnerable populations. Research in this field has the potential to positively impact the health of

individuals across the globe, and must not be forgotten in the face of short term crises.

References – Chapter 1, 2, 3, and 5

Adibi, J. J., Whyatt, R. M., Williams, P. L., Calafat, A. M., Camann, D., Herrick, R., Nelson, H., Bhat, H. K., Perera, F. P., Silva, M. J., & Hauser, R. (2008). Characterization of Phthalate Exposure among Pregnant Women Assessed by Repeat Air and Urine Samples. *Environmental Health Perspectives*, *116*(4), 467–473. https://doi.org/10.1289/ehp.10749

Andrade, A. J. M., Grande, S. W., Talsness, C. E., Grote, K., & Chahoud, I. (2006). A dose–response study following in utero and lactational exposure to di-(2-ethylhexyl)-phthalate (DEHP): Non-monotonic dose–response and low dose effects on rat brain aromatase activity. *Toxicology*, *227*(3), 185–192. https://doi.org/10.1016/J.TOX.2006.07.022

Angerer, J., Ewers, U., & Wilhelm, M. (2007). Human biomonitoring: state of the art. *International Journal of Hygiene and Environmental Health*, *210*(3–4), 201–228. https://doi.org/10.1016/j.ijheh.2007.01.024

Antonow-Schlorke, I., Schwab, M., Li, C., & Nathanielsz, P. W. (2003). Glucocorticoid exposure at the dose used clinically alters cytoskeletal proteins and presynaptic terminals in the fetal baboon brain. *Journal of Physiology*, *547*(1), 117–123. https://doi.org/10.1113/jphysiol.2002.025700

Araki, A., Mitsui, T., Goudarzi, H., Nakajima, T., Miyashita, C., Itoh, S., Sasaki, S., Cho, K., Moriya, K., Shinohara, N., Nonomura, K., & Kishi, R. (2017). Prenatal di(2-ethylhexyl) phthalate exposure and disruption of adrenal androgens and glucocorticoids levels in cord blood: The Hokkaido Study. *Science of The Total Environment*, *581*(Complete), 297–304. https://doi.org/10.1016/j.scitotenv.2016.12.124

Arbuckle, T., Davis, K., Marro, L., Fisher, M., Legrand, M., LeBlanc, A., Gaudreau, E., Foster, W. G., Choeurng, V., Fraser, W. D., & Group, M. S. (2014). Phthalate and bisphenol A exposure among pregnant women in Canada--results from the MIREC study. *Environment International*, *68*, 55–65. https://doi.org/10.1016/j.envint.2014.02.010 [doi]

Arbuckle, T., Fisher, M., MacPherson, S., Lang, C., Provencher, G., LeBlanc, A., Hauser, R., Feeley, M., Ayotte, P., Neisa, A., Ramsay, T., & Tawagi, G. (2016). Maternal and early life exposure to phthalates: The Plastics and Personal-care Products use in Pregnancy (P4) study. *Sci Total Environ*, *551–552*, 344–356. https://doi.org/10.1016/j.scitotenv.2016.02.022

Arbuckle, T E, Fraser, W. D., Fisher, M., Davis, K., Liang, C. L., Lupien, N., Bastien, S., Velez, M. P., von Dadelszen, P., Hemmings, D. G., Wang, J., Helewa, M., Taback, S., Sermer, M., Foster, W., Ross, G., Fredette, P., Smith, G., Walker, M., … Ouellet, E. (2013). Cohort profile: the maternal-infant research on environmental chemicals research platform. *Paediatric and Perinatal Epidemiology*, *27*(4), 415–425. https://doi.org/10.1111/ppe.12061 [doi]

Arbuckle, Tye E, Liang, C. L., Morisset, A.-S., Fisher, M., Weiler, H., Cirtiu, C. M., Legrand, M., Davis, K., Ettinger, A. S., & Fraser, W. D. (2016). Maternal and fetal exposure to cadmium, lead, manganese and mercury: The MIREC study. *Chemosphere*, *163*, 270–282. https://doi.org/https://doi.org/10.1016/j.chemosphere.2016.08.023

Attina, T. M., Hauser, R., Sathyanarayana, S., Hunt, P. A., Bourguignon, J.-P., Myers, J. P., DiGangi, J., Zoeller, R. T., & Trasande, L. (2016). Exposure to endocrine-disrupting chemicals in the USA: a population-based disease burden and cost analysis. *The Lancet. Diabetes & Endocrinology*, *4*(12), 996–1003. https://doi.org/10.1016/S2213-8587(16)30275-3

Autian, J. (1973). Toxicity and health threats of phthalate esters: review of the literature. *Environmental Health Perspectives, 4*, 3–26. https://doi.org/10.1289/ehp.73043

Aylward, L. L., Hays, S. M., & Zidek, A. (2017). Variation in urinary spot sample, 24 h samples, and longer-term average urinary concentrations of short-lived environmental chemicals: implications for exposure assessment and reverse dosimetry. *J Expo Sci Environ Epidemiol, 27*(6), 582–590. https://doi.org/10.1038/jes.2016.54

Bello, G. A., Arora, M., Austin, C., Horton, M. K., Wright, R. O., & Gennings, C. (2017). Extending the Distributed Lag Model framework to handle chemical mixtures. *Environmental Research, 156*, 253–264. https://doi.org/10.1016/j.envres.2017.03.031

Beltifa, A., Belaid, A., Lo Turco, V., Machreki, M., Ben Mansour, H., & Di Bella, G. (2018). Preliminary evaluation of plasticizer and BPA in Tunisian cosmetics and investigation of hazards on human skin cells. *International Journal of Environmental Health Research, 28*(5), 491–501. https://doi.org/10.1080/09603123.2018.1489528

Beltifa, A., Feriani, A., Macherki, M., Ghorbel, A., Ghazouani, L., Di Bella, G., Sire, O., Van Loco, J., Reyns, T., & Mansour, H. Ben. (2018). Persistent plasticizers and bisphenol in the cheese of Tunisian markets induced biochemical and histopathological alterations in male BALB/c mice. *Environmental Science and Pollution Research International, 25*(7), 6545–6557. https://doi.org/10.1007/s11356-017-0857-6

Bergman, Å., Heindel, J., Jobling, S., Kidd, K., & Zoeller, R. T. (2012). State-of-the-science of endocrine disrupting chemicals, 2012. In *Toxicology Letters* (Vol. 211). https://doi.org/10.1016/j.toxlet.2012.03.020

Bernal, J. (2005). Thyroid Hormones and Brain Development. In *Vitamins & Hormones* (Vol. 71, pp. 95–122). Academic Press. https://doi.org/https://doi.org/10.1016/S0083-6729(05)71004-9

Bobb, J. F., Valeri, L., Claus Henn, B., Christiani, D. C., Wright, R. O., Mazumdar, M., Godleski, J. J., & Coull, B. A. (2015). Bayesian kernel machine regression for estimating the health effects of multi-pollutant mixtures. *Biostatistics (Oxford, England), 16*(3), 493–508. https://doi.org/10.1093/biostatistics/kxu058

Bornehag, C. G., Lundgren, B., Weschler, C. J., Sigsgaard, T., Hagerhed-Engman, L., & Sundell, J. (2005). Phthalates in indoor dust and their association with building characteristics. *Environmental Health Perspectives, 113*(10), 1399–1404. https://doi.org/10.1289/ehp.7809

Bouchard, M. F., Chevrier, J., Harley, K. G., Kogut, K., Vedar, M., Calderon, N., Trujillo, C., Johnson, C., Bradman, A., Barr, D. B., & Eskenazi, B. (2011). Prenatal exposure to organophosphate pesticides and IQ in 7-year-old children. *Environ Health Perspect, 119*(8), 1189–1195.

Boyd, D. R., & Genuis, S. J. (2008). The environmental burden of disease in Canada: respiratory disease, cardiovascular disease, cancer, and congenital affliction. *Environmental Research, 106*(2), 240–249. https://doi.org/10.1016/j.envres.2007.08.009

Braaten, E. B., & Norman, D. (2006). Intelligence (IQ) Testing. *Pediatrics in Review, 27*(11), 403 LP – 408. https://doi.org/10.1542/pir.27-11-403

Braun, J., Messerlian, C., & Hauser, R. (2017). Fathers Matter: Why It's Time to Consider the Impact of

Paternal Environmental Exposures on Children's Health. *Current Epidemiology Reports, 4*(1), 46–55. https://doi.org/10.1007/s40471-017-0098-8

Braun, J M, & Gray, K. (2017). Challenges to studying the health effects of early life environmental chemical exposures on children's health. *PLoS Biol, 15*(12), e2002800.

Braun, Joseph M. (2017). Early-life exposure to EDCs: role in childhood obesity and neurodevelopment. *Nature Reviews.Endocrinology, 13*(3), 161–173. https://doi.org/10.1038/nrendo.2016.186

Braun, Joseph M, Gennings, C., Hauser, R., & Webster, T. F. (2016). What Can Epidemiological Studies Tell Us about the Impact of Chemical Mixtures on Human Health? *Environmental Health Perspectives, 124*(1), A6–A9. https://doi.org/10.1289/ehp.1510569

Buckley, J. P., Kim, H., Wong, E., & Rebholz, C. M. (2019). Ultra-processed food consumption and exposure to phthalates and bisphenols in the US National Health and Nutrition Examination Survey, 2013-2014. *Environment International, 131*, 105057. https://doi.org/10.1016/j.envint.2019.105057

Buckley, J. P., Palmieri, R. T., Matuszewski, J. M., Herring, A. H., Baird, D. D., Hartmann, K. E., & Hoppin, J. A. (2012). Consumer product exposures associated with urinary phthalate levels in pregnant women. *Journal of Exposure Science and Environmental Epidemiology, 22*(5), 468–475. https://doi.org/10.1038/jes.2012.33

Buss, C., Entringer, S., Swanson, J. M., & Wadhwa, P. D. (2012). The Role of Stress in Brain Development: The Gestational Environment's Long-Term Effects on the Brain. *Cerebrum : The Dana Forum on Brain Science, 2012*, 4. https://pubmed.ncbi.nlm.nih.gov/23447790

Calafat, A. M., Longnecker, M. P., Koch, H. M., Swan, S. H., Hauser, R., Goldman, L. R., Lanphear, B. P., Rudel, R. A., Engel, S. M., Teitelbaum, S. L., Whyatt, R. M., & Wolff, M. S. (2015). Optimal Exposure Biomarkers for Nonpersistent Chemicals in Environmental Epidemiology. *Environmental Health Perspectives, 123*(7), A166–A168. https://doi.org/10.1289/ehp.1510041

Calafat, A. M., & Needham, L. L. (2009). What additional factors beyond state-of-the-art analytical methods are needed for optimal generation and interpretation of biomonitoring data? *Environmental Health Perspectives, 117*(10), 1481–1485. https://doi.org/10.1289/ehp.0901108

Calvin, C. M., Batty, G. D., Der, G., Brett, C. E., Taylor, A., Pattie, A., Čukić, I., & Deary, I. J. (2017). Childhood intelligence in relation to major causes of death in 68 year follow-up: prospective population study. *BMJ (Clinical Research Ed.), 357*, j2708. https://doi.org/10.1136/bmj.j2708

Calvin, C. M., Deary, I. J., Fenton, C., Roberts, B. A., Der, G., Leckenby, N., & Batty, G. D. (2011). Intelligence in youth and all-cause-mortality: systematic review with meta-analysis. *International Journal of Epidemiology, 40*(3), 626–644. https://doi.org/10.1093/ije/dyq190

Canada Consumer Product Safety Act Phthalate Regulations, SOR /2016-188 (2016).

Canada, H. (2013). *Second Report on Human Biomonitoring of Environmental Chemicals in Canada: Results of the Canadian Health Measures Survey Cycle 2 (2009-2011).* Health Canada.

Cantonwine, D. E., Cordero, J. F., Rivera-González, L. O., Del, L. V. A., Ferguson, K. K., Mukherjee, B., Calafat, A. M., Crespo, N., Padilla, I. Y., Alshawabkeh, A. N., Meeker, J. D., Anzalota Del Toro, L. V,

Ferguson, K. K., Mukherjee, B., Calafat, A. M., Crespo, N., Jiménez-Vélez, B., Padilla, I. Y.,
Alshawabkeh, A. N., & Meeker, J. D. (2014). Urinary phthalate metabolite concentrations among
pregnant women in Northern Puerto Rico: distribution, temporal variability, and predictors.
Environment International, *62*, 1–11. https://doi.org/10.1016/j.envint.2013.09.014

Cao, X.-L., Sparling, M., Zhao, W., & Arbuckle, T. E. (2021). GC-MS Analysis of Phthalates and Di-(2-
thylhexyl) Adipate in Canadian Human Milk for Exposure Assessment of Infant Population. *Journal
of AOAC INTERNATIONAL*, *104*(1), 98–102. https://doi.org/10.1093/jaoacint/qsaa108

Carlin, D. J., Rider, C. V, Woychik, R., & Birnbaum, L. S. (2013). Unraveling the health effects of
environmental mixtures: an NIEHS priority. *Environmental Health Perspectives*, *121*, A6+.
http://link.galegroup.com/apps/doc/A332248578/AONE?u=winn58497&sid=AONE&xid=187e5a8b

Carlstedt, F., Jönsson, B. A. G., & Bornehag, C. G. (2013). PVC flooring is related to human uptake of
phthalates in infants. *Indoor Air*, *23*(1), 32–39. https://doi.org/10.1111/j.1600-0668.2012.00788.x

Carrico, C., Gennings, C., Wheeler, D. C., & Factor-Litvak, P. (2015). Characterization of Weighted
Quantile Sum Regression for Highly Correlated Data in a Risk Analysis Setting. *Journal of
Agricultural, Biological, and Environmental Statistics*, *20*(1), 100–120.
https://doi.org/10.1007/s13253-014-0180-3

Canadian Environmental Protection Act, 1999 Loi canadienne sur la protection de l'environnement (
1999), Organisation for Economic Co-Operation and Development (1999).
https://doi.org/ENV/JM/MONO(2007)10

Chin, H. B., Jukic, A. M., Wilcox, A. J., Weinberg, C. R., Ferguson, K. K., Calafat, A. M., McConnaughey, D.
R., & Baird, D. D. (2019). Association of urinary concentrations of early pregnancy phthalate
metabolites and bisphenol A with length of gestation. *Environmental Health*, *18*(1), 80.
https://doi.org/10.1186/s12940-019-0522-2

Hygienic standards for uses of additives in food containers and packaging materials, (2008).
https://www.chinesestandard.net/PDF/English.aspx/GB9685-2008?Redirect=YES

Chiu, Y.-H., Bellavia, A., James-Todd, T., Correia, K. F., Valeri, L., Messerlian, C., Ford, J. B., Mínguez-
Alarcón, L., Calafat, A. M., Hauser, R., Williams, P. I., & Team, E. S. (2018). Evaluating effects of
prenatal exposure to phthalate mixtures on birth weight: A comparison of three statistical
approaches. *Environment International*, *113*, 231–239.
https://doi.org/10.1016/j.envint.2018.02.005

Cho, S.-C., Bhang, S.-Y., Hong, Y.-C., Shin, M.-S., Kim, B.-N., Kim, J.-W., Yoo, H.-J., Cho, I. H., & Kim, H.-W.
(2010). Relationship between Environmental Phthalate Exposure and the Intelligence of School-
Age Children. *Environmental Health Perspectives*, *118*(7), 1027–1032.
https://doi.org/10.1289/ehp.0901376

Choi, W., Kim, S., Baek, Y.-W., Choi, K., Lee, K., Kim, S., Yu, S. Do, & Choi, K. (2017). Exposure to
environmental chemicals among Korean adults-updates from the second Korean National
Environmental Health Survey (2012–2014). *International Journal of Hygiene and Environmental
Health*, *220*(2, Part A), 29–35. https://doi.org/https://doi.org/10.1016/j.ijheh.2016.10.002

Christen, V., Crettaz, P., Oberli-Schrämmli, A., & Fent, K. (2012). Antiandrogenic activity of phthalate
mixtures: Validity of concentration addition. *Toxicology and Applied Pharmacology*, *259*(2), 169–

176. https://doi.org/https://doi.org/10.1016/j.taap.2011.12.021

Colborn, T., vom Saal, F. S., & Soto, A. M. (1993). Developmental effects of endocrine-disrupting chemicals in wildlife and humans. *Environmental Health Perspectives, 101*(5), 378–384. https://doi.org/10.1289/ehp.93101378

Colciago, A., Negri-Cesi, P., Pravettoni, A., Mornati, O., Casati, L., & Celotti, F. (2006). Prenatal Aroclor 1254 exposure and brain sexual differentiation: Effect on the expression of testosterone metabolizing enzymes and androgen receptors in the hypothalamus of male and female rats. *Reproductive Toxicology, 22*(4), 738–745. https://doi.org/10.1016/j.reprotox.2006.07.002

Colicino, E., Foppa Pedretti, N., Busgang, S., & Gennings, C. (2019). Per- and poly-fluoroalkyl substances and bone mineral density: results from the Bayesian weighted quantile sum regression. *MedRxiv,* 19010710. https://doi.org/10.1101/19010710

Consumer product safety improvement act of 2008, 153 (2008).

Daniel, S., Balalian, A. A., Whyatt, R. M., Liu, X., Rauh, V., Herbstman, J., & Factor-Litvak, P. (2020). Perinatal phthalates exposure decreases fine-motor functions in 11-year-old girls: Results from weighted Quantile sum regression. *Environment International, 136*, 105424. https://doi.org/https://doi.org/10.1016/j.envint.2019.105424

Desrochers-Couture, M., Oulhote, Y., Arbuckle, T. E., Fraser, W. D., Seguin, J. R., Ouellet, E., Forget-Dubois, N., Ayotte, P., Boivin, M., Lanphear, B. P., & Muckle, G. (2018). Prenatal, concurrent, and sex-specific associations between blood lead concentrations and IQ in preschool Canadian children. *Environment International, 121*(Pt 2), 1235–1242. https://doi.org/10.1016/j.envint.2018.10.043

Diamanti-Kandarakis, E., Bourguignon, J. P., Giudice, L. C., Hauser, R., Prins, G. S., Soto, A. M., Zoeller, R. T., & Gore, A. C. (2009). Endocrine-disrupting chemicals: an Endocrine Society scientific statement. *Endocrine Reviews, 30*(4), 293–342. https://doi.org/10.1210/er.2009-0002 [doi]

Dixon, H. M., Armstrong, G., Barton, M., Bergmann, A. J., Bondy, M., Halbleib, M. L., Hamilton, W., Haynes, E., Herbstman, J., Hoffman, P., Jepson, P., Kile, M. L., Kincl, L., Laurienti, P. J., North, P., Paulik, L. B., Petrosino, J., Points, G. L. 3rd, Poutasse, C. M., … Anderson, K. A. (2019). Discovery of common chemical exposures across three continents using silicone wristbands. *Royal Society Open Science, 6*(2), 181836. https://doi.org/10.1098/rsos.181836

Dobson, K. G., Chow, C. H. T., Morrison, K. M., & Van Lieshout, R. J. (2017). Associations Between Childhood Cognition and Cardiovascular Events in Adulthood: A Systematic Review and Meta-analysis. *The Canadian Journal of Cardiology, 33*(2), 232–242. https://doi.org/10.1016/j.cjca.2016.08.014

Doherty, B. T., Engel, S. M., Buckley, J. P., Silva, M. J., Calafat, A. M., & Wolff, M. S. (2017). Prenatal Phthalate Biomarker Concentrations and Performance on the Bayley Scales of Infant Development-II in a Population of Young Urban Children. *Environmental Research, 152*, 51–58. https://doi.org/10.1016/j.envres.2016.09.021

Dong, R., Wu, Y., Chen, J., Wu, M., Li, S., & Chen, B. (2019). Lactational exposure to phthalates impaired the neurodevelopmental function of infants at 9 months in a pilot prospective study. *Chemosphere, 226*, 351–359. https://doi.org/https://doi.org/10.1016/j.chemosphere.2019.03.159

Dong, X., Dong, J., Zhao, Y., Guo, J., Wang, Z., Liu, M., Zhang, Y., & Na, X. (2017). Effects of long-term in vivo exposure to di-2-ethylhexylphthalate on thyroid hormones and the TSH/TSHR signaling pathways in wistar rats. *International Journal of Environmental Research and Public Health, 14*, 44. https://doi.org/10.3390/ijerph14010044

EFSA Panel on Contaminants in the Food Chain. (2010). Scientific Opinion on Lead in Food. In *EFSA Journal* (Vol. 8, Issue 4). John Wiley & Sons, Ltd. https://doi.org/https://doi.org/10.2903/j.efsa.2010.1570

Ellington, J. J. (1999). Octanol/Water Partition Coefficients and Water Solubilities of Phthalate Esters. *Journal of Chemical & Engineering Data, 44*(6), 1414–1418. https://doi.org/10.1021/je990149u

Engel, S. M., Miodovnik, A., Canfield, R. L., Zhu, C., Silva, M. J., Calafat, A. M., & Wolff, M. S. (2010). Prenatal phthalate exposure is associated with childhood behavior and executive functioning. *Environ Health Perspect, 118*(4), 565–571.

Engel, S. M., Villanger, G. D., Nethery, R. C., Thomsen, C., Sakhi, A. K., Drover, S. S. M., Hoppin, J. A., Zeiner, P., Knudsen, G. P., Reichborn-Kjennerud, T., Herring, A. H., & Aase, H. (2018). Prenatal Phthalates, Maternal Thyroid Function, and Risk of Attention-Deficit Hyperactivity Disorder in the Norwegian Mother and Child Cohort. In *Environ Health Perspect* (Vol. 126, p. 57004). https://doi.org/10.1289/ehp2358

Environment and Climate Change Canada, & Canada, H. (2008). *Screening Assessment for the Challenge (Bisphenol A)* (Issue 80).

Environment and Climate Change Canada, & Health Canada. (2015a). *State of the Science: Report Phthalate Substance Grouping - Medium-Chain Phthalate Esters.* https://www.ec.gc.ca/ese-ees/4D845198-761D-428B-A519-75481B25B3E5/SoS_Phthalates(Medium-chain)_EN.pdf

Environment and Climate Change Canada, & Health Canada. (2015b). *State of the Science Report Phthalate Substance Grouping Long-chain Phthalate Esters* (Issue August).

Environment Canada, H. C. (2017). *Draft Screening Assessment Phthalate Substance Grouping.* https://www.ec.gc.ca/ese-ees/516A504A-0A21-4AF5-8310-ADD2FE5C0C76/DSAR Phthalates-EN.pdf

Environmental Protection Agency. (2017). *America's Children and the Environment - Biomonitoring Phthalates* (Issue August). https://www.epa.gov/sites/production/files/2017-08/documents/phthalates_updates_live_file_508_0.pdf

European Chemicals Agency, E. (2017). *Inclusion of substances of very high concern in the Candidate List for eventual inclusion in Annex XIV* (Vol. 1, Issue 5).

European Commission. (1999). *adopting measures prohibiting the placing on the market of toys and childcare articles intended to be placed in the mouth by children under three years of age made of soft PVC containing one or more of the substances di-iso-nonyl phthalate (DINP), di(2-et. 8*, 46–49. http://eur-lex.europa.eu/legal-content/EN/TXT/PDF/?uri=CELEX:31999D0815&qid=1453399198035&from=EN

European Parliament. (2005). Directive 2005/84/EC of the European Parliament and the Council. *Official Journal of the European Union, 344*(40), 40–43.

Evans, D., Chaix, B., Lobbedez, T., Verger, C., & Flahault, A. (2012). Combining directed acyclic graphs and the change-in-estimate procedure as a novel approach to adjustment-variable selection in epidemiology. *BMC Medical Research Methodology, 12*(1), 156. https://doi.org/10.1186/1471-2288-12-156

Factor-Litvak, P., Insel, B., Calafat, A., Liu, X., Perera, F., Rauh, V., & Whyatt, R. (2014). Persistent Associations between Maternal Prenatal Exposure to Phthalates on Child IQ at Age 7 Years: e114003. *PLoS ONE, 9*(12). https://doi.org/10.1371/journal.pone.0114003

Ferguson, K. K., McElrath, T. F., & Meeker, J. D. (2014). Environmental phthalate exposure and preterm birth. *JAMA Pediatrics, 168*(1), 61–67. https://doi.org/10.1001/jamapediatrics.2013.3699

Fierens, T., Servaes, K., Van Holderbeke, M., Geerts, L., De Henauw, S., Sioen, I., & Vanermen, G. (2012). Analysis of phthalates in food products and packaging materials sold on the Belgian market. *Food and Chemical Toxicology.* https://doi.org/10.1016/j.fct.2012.04.029

Fisher, M., Arbuckle, T. E., MacPherson, S., Braun, J. M., Feeley, M., & Gaudreau, E. (2019). Phthalate and BPA Exposure in Women and Newborns through Personal Care Product Use and Food Packaging. *Environmental Science & Technology, 53*(18), 10813–10826. https://doi.org/10.1021/acs.est.9b02372

Fisher, M., Arbuckle, T. E., Mallick, R., LeBlanc, A., Hauser, R., Feeley, M., Koniecki, D., Ramsay, T., Provencher, G., Bérubé, R., & Walker, M. (2015). Bisphenol A and phthalate metabolite urinary concentrations: Daily and across pregnancy variability. *Journal of Exposure Science & Environmental Epidemiology, 25*(3), 231–239. http://10.0.4.14/jes.2014.65

Forns, J., Aranbarri, A., Grellier, J., Julvez, J., Vrijheid, M., & Sunyer, J. (2012). A Conceptual Framework in the Study of Neuropsychological Development in Epidemiological Studies. *Neuroepidemiology, 38*(4), 203–208. https://doi.org/10.1159/000337169

Foster, P. M. D. (2005). Mode of action: impaired fetal leydig cell function--effects on male reproductive development produced by certain phthalate esters. *Critical Reviews in Toxicology, 35*(8–9), 713–719. https://doi.org/10.1080/10408440591007395

Frederiksen, H., Jorgensen, N., & Andersson, A.-M. (2010). Correlations between phthalate metabolites in urine, serum, and seminal plasma from young Danish men determined by isotope dilution liquid chromatography tandem mass spectrometry. *Journal of Analytical Toxicology, 34*(7), 400–410. https://doi.org/10.1093/jat/34.7.400

Frederiksen, H., Nielsen, O., Koch, H. M., Skakkebaek, N. E., Juul, A., Jørgensen, N., & Andersson, A.-M. (2020). Changes in urinary excretion of phthalates, phthalate substitutes, bisphenols and other polychlorinated and phenolic substances in young Danish men; 2009–2017. *International Journal of Hygiene and Environmental Health, 223*(1), 93–105. https://doi.org/https://doi.org/10.1016/j.ijheh.2019.10.002

Fromme, H., Gruber, L., Seckin, E., Raab, U., Zimmermann, S., Kiranoglu, M., Schlummer, M., Schwegler, U., Smolic, S., & Volkel, W. (2011). Phthalates and their metabolites in breast milk--results from the Bavarian Monitoring of Breast Milk (BAMBI). *Environ Int, 37*(4), 715–722. https://doi.org/10.1016/j.envint.2011.02.008

Gale, C. R., O'Callaghan, F. J., Bredow, M., & Martyn, C. N. (2006). The influence of head growth in fetal life, infancy, and childhood on intelligence at the ages of 4 and 8 years. *Pediatrics, 118*(4), 1486–1492. https://doi.org/10.1542/peds.2005-2629

Gao, H., Wu, W., Xu, Y., Jin, Z., Bao, H., Zhu, P., Su, P., Sheng, J., Hao, J., & Tao, F. (2017). Effects of Prenatal Phthalate Exposure on Thyroid Hormone Concentrations Beginning at The Embryonic Stage. *Scientific Reports, 7*, 13106. https://doi.org/10.1038/s41598-017-13672-x

Gaylord, A., Osborne, G., Ghassabian, A., Malits, J., Attina, T., & Trasande, L. (2020). Trends in neurodevelopmental disability burden due to early life chemical exposure in the USA from 2001 to 2016: A population-based disease burden and cost analysis. *Molecular and Cellular Endocrinology, 502*, 110666. https://doi.org/https://doi.org/10.1016/j.mce.2019.110666

Ghisari, M., & Bonefeld Jorgensen, E. C. (2009). Effects of plasticizers and their mixtures on estrogen receptor and thyroid hormone functions. *Toxicology Letters, 189*(1), 67–77. https://doi.org/https://doi.org/10.1016/j.toxlet.2009.05.004

Gordon, B. (2004). Test Review: Wechsler, D. (2002). The Wechsler Preschool and Primary Scale of Intelligence, Third Edition (WPPSI-III). San Antonio, TX: The Psychological Corporation. *Canadian Journal of School Psychology, 19*(1–2), 205–220. https://doi.org/10.1177/082957350401900111

Gore, A. C., Chappell, V. A., Fenton, S. E., Flaws, J. A., Nadal, A., Prins, G. S., Toppari, J., & Zoeller, R. T. (2015). EDC-2: The Endocrine Society's Second Scientific Statement on Endocrine-Disrupting Chemicals. *Endocrine Reviews, 36*(6), E1–E150. https://doi.org/10.1210/er.2015-1010

Graham, P. R. (1973). Phthalate ester plasticizers--why and how they are used. *Environmental Health Perspectives, 3*, 3–12. https://doi.org/10.1289/ehp.73033

Grandjean, P., & Landrigan, P. J. (2006). Developmental neurotoxicity of industrial chemicals. *Lancet (London, England), 368*(9553), 2167–2178. https://doi.org/10.1016/S0140-6736(06)69665-7

Grandjean, P., & Landrigan, P. J. (2014). Neurobehavioural effects of developmental toxicity. *Lancet Neurol, 13*(3), 330–338.

Grandjean, P., Weihe, P., White, R. F., Debes, F., Araki, S., Yokoyama, K., Murata, K., Sorensen, N., Dahl, R., & Jorgensen, P. J. (1997). Cognitive deficit in 7-year-old children with prenatal exposure to methylmercury. In *Neurotoxicol Teratol* (Vol. 19, pp. 417–428).

Gray, T. J., & Gangolli, S. D. (1986). Aspects of the testicular toxicity of phthalate esters. *Environmental Health Perspectives, 65*, 229–235. https://doi.org/10.1289/ehp.8665229

Greenland, S. (1989). Modeling and variable selection in epidemiologic analysis. *American Journal of Public Health, 79*(3), 340–349. https://doi.org/10.2105/ajph.79.3.340

Greenland, Sander, Daniel, R., & Pearce, N. (2016). Outcome modelling strategies in epidemiology: traditional methods and basic alternatives. *International Journal of Epidemiology, 45*(2), 565–575. https://doi.org/10.1093/ije/dyw040

Greenland, Sander, & Pearce, N. (2015). Statistical Foundations for Model-Based Adjustments. *Annual Review of Public Health, 36*(1), 89–108. https://doi.org/10.1146/annurev-publhealth-031914-122559

Greenland, Sander, Pearl, J., & Robins, J. M. (1999). Causal Diagrams for Epidemiologic Research. *Epidemiology, 10*(1). https://journals.lww.com/epidem/Fulltext/1999/01000/Causal_Diagrams_for_Epidemiologic_Rese arch.8.aspx

Guo, Y., Alomirah, H., Cho, H.-S., Minh, T. B., Mohd, M. A., Nakata, H., & Kannan, K. (2011). Occurrence of Phthalate Metabolites in Human Urine from Several Asian Countries. *Environmental Science & Technology, 45*(7), 3138–3144. https://doi.org/10.1021/es103879m

Gyllenhammar, I., Glynn, A., Jonsson, B. A. G., Lindh, C. H., Darnerud, P. O., Svensson, K., & Lignell, S. (2017). Diverging temporal trends of human exposure to bisphenols and plastizisers, such as phthalates, caused by substitution of legacy EDCs? *Environmental Research, 153*, 48–54. https://doi.org/10.1016/j.envres.2016.11.012

Haddow, J. E., Palomaki, G. E., Allan, W. C., Williams, J. R., Knight, G. J., Gagnon, J., O'Heir, C. E., Mitchell, M. L., Hermos, R. J., Waisbren, S. E., Faix, J. D., & Klein, R. Z. (1999). Maternal Thyroid Deficiency during Pregnancy and Subsequent Neuropsychological Development of the Child. *New England Journal of Medicine, 341*(8), 549–555. https://doi.org/10.1056/NEJM199908193410801

Haines, D. A., Saravanabhavan, G., Werry, K., & Khoury, C. (2017). An overview of human biomonitoring of environmental chemicals in the Canadian Health Measures Survey: 2007–2019. *International Journal of Hygiene and Environmental Health, 220*(2, Part A), 13–28. https://doi.org/https://doi.org/10.1016/j.ijheh.2016.08.002

Hammel, S. C., Levasseur, J. L., Hoffman, K., Phillips, A. L., Lorenzo, A. M., Calafat, A. M., Webster, T. F., & Stapleton, H. M. (2019). Children's exposure to phthalates and non-phthalate plasticizers in the home: The TESIE study. *Environment International, 132*, 105061. https://doi.org/https://doi.org/10.1016/j.envint.2019.105061

Hannas, B. R., Lambright, C. S., Furr, J., Howdeshell, K. L., Wilson, V. S., & Gray Jr, L. E. (2011). Dose-Response Assessment of Fetal Testosterone Production and Gene Expression Levels in Rat Testes Following InUtero Exposure to Diethylhexyl Phthalate, Diisobutyl Phthalate, Diisoheptyl Phthalate, and Diisononyl Phthalate. *Toxicological Sciences, 123*(1), 206–216. https://doi.org/10.1093/toxsci/kfr146

Haugen, A. C., Schug, T. T., Collman, G., & Heindel, J. J. (2015). Evolution of DOHaD: the impact of environmental health sciences. In *Journal of developmental origins of health and disease* (Vol. 6, Issue 2, pp. 55–64). https://doi.org/10.1017/S2040174414000580

Health Canada. (2010). *Report on Human Biomonitoring of Environmental Chemicals in Canada* (Vol. 1). https://www.canada.ca/en/health-canada/services/environmental-workplace-health/reports-publications/environmental-contaminants/report-human-biomonitoring-environmental-chemicals-canada-health-canada-2010.html

Health Canada. (2017). *Fourth Report on Human Biomonitoring of Environmental Chemicals in Canada: Results of the Canadian Health Measures Survey Cycle 4 (2014-2015)*. Health Canada.

Health Canada. (2019). *FIFTH REPORT ON HUMAN BIOMONITORING OF ENVIRONMENTAL CHEMICALS IN CANADA: Results of the Canadian Health Measures Survey Cycle 5 (2016–2017)*. https://www.canada.ca/content/dam/hc-sc/documents/services/environmental-workplace-health/reports-publications/environmental-contaminants/fifth-report-human-biomonitoring/pub1-eng.pdf

Henrichs, J., Ghassabian, A., Peeters, R. P., & Tiemeier, H. (2013). Maternal hypothyroxinemia and effects on cognitive functioning in childhood: How and why? *Clinical Endocrinology*, *79*(2), 152–162. https://doi.org/10.1111/cen.12227

Heudorf, U., Mersch-Sundermann, V., & Angerer, J. (2007). Phthalates: Toxicology and exposure. *International Journal of Hygiene and Environmental Health*, *210*(5), 623–634. https://doi.org/10.1016/j.ijheh.2007.07.011

Hong, D., Li, X.-W., Lian, Q.-Q., Lamba, P., Bernard, D. J., Hardy, D. O., Chen, H.-X., & Ge, R.-S. (2009). Mono-(2-ethylhexyl) phthalate (MEHP) regulates glucocorticoid metabolism through 11β-hydroxysteroid dehydrogenase 2 in murine gonadotrope cells. *Biochemical and Biophysical Research Communications*, *389*(2), 305–309. https://doi.org/https://doi.org/10.1016/j.bbrc.2009.08.134

Howdeshell, K. L., Wilson, V. S., Furr, J., Lambright, C. R., Rider, C. V, Blystone, C. R., Hotchkiss, A. K., & Gray Jr, L. E. (2008). A Mixture of Five Phthalate Esters Inhibits Fetal Testicular Testosterone Production in the Sprague-Dawley Rat in a Cumulative, Dose-Additive Manner. *Toxicological Sciences*, *105*(1), 153–165. https://doi.org/10.1093/toxsci/kfn077

Huang, H. B., Chen, H. Y., Su, P. H., Huang, P. C., Sun, C. W., Wang, C. J., Hsiung, C. A., & Wang, S. L. (2015). Fetal and Childhood Exposure to Phthalate Diesters and Cognitive Function in Children Up to 12 Years of Age: Taiwanese Maternal and Infant Cohort Study. In A. Gonzalez-Bulnes (Ed.), *PLoS One* (Vol. 10). https://doi.org/10.1371/journal.pone.0131910

Huang, P. C., Tsai, C. H., Liang, W. Y., Li, S. S., Huang, H. B., & Kuo, P. L. (2016). Early Phthalates Exposure in Pregnant Women Is Associated with Alteration of Thyroid Hormones. *PLoS One*, *11*(7), e0159398.

Husøy, T., Andreassen, M., Hjertholm, H., Carlsen, M. H., Norberg, N., Sprong, C., Papadopoulou, E., Sakhi, A. K., Sabaredzovic, A., & Dirven, H. A. A. M. (2019). The Norwegian biomonitoring study from the EU project EuroMix: Levels of phenols and phthalates in 24-hour urine samples and exposure sources from food and personal care products. *Environment International*. https://doi.org/10.1016/j.envint.2019.105103

Hyland, C., Mora, A. M., Kogut, K., Calafat, A. M., Harley, K., Deardorff, J., Holland, N., Eskenazi, B., & Sagiv, S. K. (2019). Prenatal Exposure to Phthalates and Neurodevelopment in the CHAMACOS Cohort. *Environmental Health Perspectives*, *127*(10), 107010. https://doi.org/10.1289/EHP5165

Jacobson, J. L., & Jacobson, S. W. (1996). Intellectual impairment in children exposed to polychlorinated biphenyls in utero. *The New England Journal of Medicine*, *335*(11), 783–789. https://doi.org/10.1056/NEJM199609123351104

Jaeger, R. J., & Rubin, R. J. (1970). Plasticizers from Plastic Devices: Extraction, Metabolism, and Accumulation by Biological Systems. *Science*, *170*(3956), 460 LP – 462. https://doi.org/10.1126/science.170.3956.460

Jankowska, A., Polańska, K., Koch, H. M., Pälmke, C., Waszkowska, M., Stańczak, A., Wesołowska, E., Hanke, W., Bose-O'Reilly, S., Calamandrei, G., & Garí, M. (2019). Phthalate exposure and neurodevelopmental outcomes in early school age children from Poland. *Environmental Research*, *179*, 108829. https://doi.org/10.1016/J.ENVRES.2019.108829

Jensen, M. S., Anand-Ivell, R., Nørgaard-Pedersen, B., Jönsson, B. A. G., Bonde, J. P., Hougaard, D. M., Cohen, A., Lindh, C. H., Ivell, R., & Toft, G. (2015). Amniotic fluid phthalate levels and male fetal gonad function. *Epidemiology*, *26*(1), 91–99. https://doi.org/10.1097/EDE.0000000000000198

Jianlong, W., Lujun, C., Hanchang, S., & Yi, Q. (2000). Microbial degradation of phthalic acid esters under anaerobic digestion of sludge. *Chemosphere*, *41*(8), 1245–1248. https://doi.org/10.1016/S0045-6535(99)00552-4

Johns, L E, Cooper, G. S., Galizia, A., & Meeker, J. D. (2015). Exposure Assessment Issues in Epidemiology Studies of Phthalates. *Environ Int*, *85*, 27–39.

Johns, Lauren E, Ferguson, K. K., Soldin, O. P., Cantonwine, D. E., Rivera-González, L. O., Del Toro, L. V. A., Calafat, A. M., Ye, X., Alshawabkeh, A. N., Cordero, J. F., & Meeker, J. D. (2015). Urinary phthalate metabolites in relation to maternal serum thyroid and sex hormone levels during pregnancy: a longitudinal analysis. *Reproductive Biology and Endocrinology : RB&E*, *13*, 4. https://doi.org/10.1186/1477-7827-13-4

Jolliffe, I. T. (2002). *Principal Component Analysis* (Springerlink (ed.); Second Edi). Springer New York. https://doi.org/10.1007/b98835

Joubert, B. R., Mantooth, S. N., & McAllister, K. A. (2020). Environmental Health Research in Africa: Important Progress and Promising Opportunities. *Frontiers in Genetics*, *10*, 1166. https://doi.org/10.3389/fgene.2019.01166

Jurewicz, J., & Hanke, W. (2011). Exposure to phthalates: Reproductive outcome and children health. A review of epidemiological studies. *International Journal of Occupational Medicine and Environmental Health*, *24*(2), 115–141. https://doi.org/10.2478/s13382-011-0022-2

Just, A. C., Adibi, J. J., Rundle, A. G., Calafat, A. M., Camann, D. E., Hauser, R., Silva, M. J., & Whyatt, R. M. (2010). Urinary and air phthalate concentrations and self-reported use of personal care products among minority pregnant women in New York city. *Journal of Exposure Science and Environmental Epidemiology*, *20*(7), 625–633. https://doi.org/10.1038/jes.2010.13

Just, A. C., Miller, R. L., Perzanowski, M. S., Rundle, A. G., Chen, Q., Jung, K. H., Hoepner, L., Camann, D. E., Calafat, A. M., Perera, F. P., & Whyatt, R. M. (2015). Vinyl flooring in the home is associated with children's airborne butylbenzyl phthalate and urinary metabolite concentrations. *Journal of Exposure Science & Environmental Epidemiology*, *25*(6), 574–579. https://doi.org/10.1038/jes.2015.4

Kalloo, G., Wellenius, G. A., McCandless, L., Calafat, A. M., Sjodin, A., Karagas, M., Chen, A., Yolton, K., Lanphear, B. P., & Braun, J. M. (2018). Profiles and Predictors of Environmental Chemical Mixture Exposure among Pregnant Women: The Health Outcomes and Measures of the Environment Study. *Environmental Science & Technology*, *52*(17), 10104–10113. https://doi.org/10.1021/acs.est.8b02946

Kasper-Sonnenberg, M., Koch, H. M., Apel, P., Rüther, M., Pälmke, C., Brüning, T., & Kolossa-Gehring, M. (2019). Time trend of exposure to the phthalate plasticizer substitute DINCH in Germany from 1999 to 2017: Biomonitoring data on young adults from the Environmental Specimen Bank (ESB).

International Journal of Hygiene and Environmental Health, 222(8), 1084–1092. https://doi.org/https://doi.org/10.1016/j.ijheh.2019.07.011

Keil, A., Buckley, J., O'Brien, K., Ferguson, K., Shanshan, Z., & White, A. (2020). A Quantile-Based g-Computation Approach to Addressing the Effects of Exposure Mixtures. *Environmental Health Perspectives, 128*(4), 47004. https://doi.org/10.1289/EHP5838

Keil, A., J, B., K, O., K, F., S, Z., & White, A. (2019). A quantile-based g-computation approach to addressing the effects of exposure mixtures. *Environmental Epidemiology, 3*, 44. https://doi.org/10.1097/01.EE9.0000606120.58494.9d

Kim, J. I., Hong, Y. C., Shin, C. H., Lee, Y. A., Lim, Y. H., & Kim, B. N. (2017). The effects of maternal and children phthalate exposure on the neurocognitive function of 6-year-old children. *Environmental Research, 156*, 519–525. https://doi.org/10.1016/j.envres.2017.04.003

Kim, K.-N., Kim, H. Y., Lim, Y.-H., Shin, C. H., Kim, J. I., Kim, B.-N., Lee, Y. A., & Hong, Y.-C. (2020). Prenatal and early childhood phthalate exposures and thyroid function among school-age children. *Environment International, 141*, 105782. https://doi.org/https://doi.org/10.1016/j.envint.2020.105782

Kim, S., Eom, S., Kim, H. J., Lee, J. J., Choi, G., Choi, S., Kim, S. Y., Cho, G., Kim, Y. D., Suh, E., Kim, S. K., Kim, G. H., Moon, H. B., Park, J., Choi, K., & Eun, S. H. (2018). Association between maternal exposure to major phthalates, heavy metals, and persistent organic pollutants, and the neurodevelopmental performances of their children at 1 to 2 years of age-CHECK cohort study. *Sci Total Environ, 624*, 377–384. https://doi.org/10.1016/j.scitotenv.2017.12.058

Kim, Y., Ha, E. H., Kim, E. J., Park, H., Ha, M., Kim, J. H., Hong, Y. C., Chang, N., & Kim, B. N. (2011). Prenatal exposure to phthalates and infant development at 6 months: prospective Mothers and Children's Environmental Health (MOCEH) study. *Environ Health Perspect, 119*(10), 1495–1500.

Kiralan, M., Toptanci, I., Yavuz, M., & Ramadan, M. F. (2019). Phthalates levels in cold-pressed oils marketed in Turkey. *Environmental Science and Pollution Research International*. https://doi.org/10.1007/s11356-019-07162-y

Koch, H. M., Bolt, H. M., Preuss, R., & Angerer, J. (2005). New metabolites of di(2-ethylhexyl)phthalate (DEHP) in human urine and serum after single oral doses of deuterium-labelled DEHP. *Archives of Toxicology, 79*(7), 367–376. https://doi.org/10.1007/s00204-004-0642-4

Koch, H. M., Gonzalez-Reche, L. M., & Angerer, J. (2003). On-line clean-up by multidimensional liquid chromatography–electrospray ionization tandem mass spectrometry for high throughput quantification of primary and secondary phthalate metabolites in human urine. *Journal of Chromatography B, 784*(1), 169–182. https://doi.org/https://doi.org/10.1016/S1570-0232(02)00785-7

Koniecki, D., Wang, R., Moody, R. P., & Zhu, J. (2011). Phthalates in cosmetic and personal care products: Concentrations and possible dermal exposure. *Environmental Research, 111*(3), 329–336. https://doi.org/https://doi.org/10.1016/j.envres.2011.01.013

Korevaar, T. I. M., Muetzel, R., Medici, M., Chaker, L., Jaddoe, V. W. V, de Rijke, Y. B., Steegers, E. A. P., Visser, T. J., White, T., Tiemeier, H., & Peeters, R. P. (2016). Association of maternal thyroid function during early pregnancy with offspring IQ and brain morphology in childhood: a population-based prospective cohort study. *The Lancet. Diabetes & Endocrinology, 4*(1), 35–43. https://doi.org/10.1016/S2213-8587(15)00327-7

Kortenkamp, A. (2014). Low dose mixture effects of endocrine disrupters and their implications for regulatory thresholds in chemical risk assessment. *Current Opinion in Pharmacology, 19*, 105–111. https://doi.org/10.1016/j.coph.2014.08.006

Kostović, I., Judas, M., Rados, M., & Hrabac, P. (2002). Laminar organization of the human fetal cerebrum revealed by histochemical markers and magnetic resonance imaging. *Cerebral Cortex (New York, N.Y. : 1991), 12*(5), 536–544. https://doi.org/10.1093/cercor/12.5.536

Lam, J., Lanphear, B. P., Bellinger, D., Axelrad, D. A., McPartland, J., Sutton, P., Davidson, L., Daniels, N., Sen, S., & Woodruff, T. J. (2017). Developmental PBDE Exposure and IQ/ADHD in Childhood: A Systematic Review and Meta-analysis. *Environ Health Perspect, 125*(1552-9924 (Electronic)).

Lange, N., Froimowitz, M. P., Bigler, E. D., Lainhart, J. E., & Group, B. D. C. (2010). Associations between IQ, total and regional brain volumes, and demography in a large normative sample of healthy children and adolescents. *Developmental Neuropsychology, 35*(3), 296–317. https://doi.org/10.1080/87565641003696833

Lanphear, B P, Hornung, R., Ho, M., Howard, C. R., Eberly, S., & Knauf, K. (2002). Environmental lead exposure during early childhood. In *J Pediatr* (Vol. 140, pp. 40–47).

Lanphear, Bruce P, Hornung, R., Khoury, J., Yolton, K., Baghurst, P., Bellinger, D. C., Canfield, R. L., Dietrich, K. N., Bornschein, R., Greene, T., Rothenberg, S. J., Needleman, H. L., Schnaas, L., Wasserman, G., Graziano, J., & Roberts, R. (2005). Low-level environmental lead exposure and children's intellectual function: an international pooled analysis. *Environmental Health Perspectives, 113*(7), 894–899. https://doi.org/10.1289/ehp.7688

Laskey, J. W., & Berman, E. (1993). Steroidogenic assessment using ovary culture in cycling rats: effects of bis(2-diethylhexyl)phthalate on ovarian steroid production. *Reproductive Toxicology (Elmsford, N.Y.), 7*(1), 25–33. https://doi.org/10.1016/0890-6238(93)90006-s

Lazarevic, N., Barnett, A. G., Sly, P. D., & Knibbs, L. D. (2019). Statistical methodology in studies of prenatal exposure to mixtures of endocrine-disrupting chemicals: A review of existing approaches and new alternatives. *Environmental Health Perspectives, 127*(2). https://doi.org/10.1289/EHP2207

Lee, M., Rahbar, M. H., Samms-Vaughan, M., Bressler, J., Bach, M. A., Hessabi, M., Grove, M. L., Shakespeare-Pellington, S., Coore Desai, C., Reece, J.-A., Loveland, K. A., & Boerwinkle, E. (2019). A generalized weighted quantile sum approach for analyzing correlated data in the presence of interactions. *Biometrical Journal. Biometrische Zeitschrift, 61*(4), 934–954. https://doi.org/10.1002/bimj.201800259

Lee, W.-C., Fisher, M., Davis, K., Arbuckle, T. E., & Sinha, S. K. (2017). Identification of chemical mixtures to which Canadian pregnant women are exposed: The MIREC Study. *Environment International, 99*, 321–330. https://doi.org/https://doi.org/10.1016/j.envint.2016.12.015

Li, N., Papandonatos, G. D., Calafat, A. M., Yolton, K., Lanphear, B. P., Chen, A., & Braun, J. M. (2019). Identifying periods of susceptibility to the impact of phthalates on children's cognitive abilities. *Environmental Research, 172*, 604–614.

https://doi.org/https://doi.org/10.1016/j.envres.2019.03.009

Liang, D. W., Zhang, T., Fang, H. H. P., & He, J. (2008). Phthalates biodegradation in the environment. *Applied Microbiology and Biotechnology*, *80*(2), 183–198. https://doi.org/10.1007/s00253-008-1548-5

Lioy, P. J., Hauser, R., Gennings, C., Koch, H. M., Mirkes, P. E., Schwetz, B. A., & Kortenkamp, A. (2015). Assessment of phthalates/phthalate alternatives in children's toys and childcare articles: Review of the report including conclusions and recommendation of the Chronic Hazard Advisory Panel of the Consumer Product Safety Commission. *Journal of Exposure Science and Environmental Epidemiology*, *25*(4), 343–353. https://doi.org/10.1038/jes.2015.33

Lloyd, S. C., & Foster, P. M. (1988). Effect of mono-(2-ethylhexyl)phthalate on follicle-stimulating hormone responsiveness of cultured rat Sertoli cells. *Toxicology and Applied Pharmacology*, *95*(3), 484–489. https://doi.org/10.1016/0041-008x(88)90366-3

Marcel, Y. L., & Noel, S. P. (1970). A plasticizer in lipid extracts of human blood. *Chemistry and Physics of Lipids*, *4*(3), 417–418. https://doi.org/10.1016/0009-3084(70)90040-x

Martin, L. T., & Kubzansky, L. D. (2005). Childhood cognitive performance and risk of mortality: a prospective cohort study of gifted individuals. *American Journal of Epidemiology*, *162*(9), 887–890. https://doi.org/10.1093/aje/kwi300

Matthews, L. G., Inder, T. E., Pascoe, L., Kapur, K., Lee, K. J., Monson, B. B., Doyle, L. W., Thompson, D. K., & Anderson, P. J. (2018). Longitudinal Preterm Cerebellar Volume: Perinatal and Neurodevelopmental Outcome Associations. *Cerebellum (London, England)*, *17*(5), 610–627. https://doi.org/10.1007/s12311-018-0946-1

Meeker, J. D. (2012). Exposure to environmental endocrine disruptors and child development. *Archives of Pediatrics & Adolescent Medicine*, *166*(6), E1-7. https://doi.org/10.1001/archpediatrics.2012.241

Meeker, J. D., Sathyanarayana, S., & Swan, S. H. (2009). Phthalates and other additives in plastics: human exposure and associated health outcomes. *Philosophical Transactions of the Royal Society B: Biological Sciences*, *364*(1526), 2097. http://rstb.royalsocietypublishing.org/content/364/1526/2097.abstract

Mickey, R. M., & Greenland, S. (1989). THE IMPACT OF CONFOUNDER SELECTION CRITERIA ON EFFECT ESTIMATION. *American Journal of Epidemiology*, *129*(1), 125–137. http://dx.doi.org/10.1093/oxfordjournals.aje.a115101

Miodovnik, A., Edwards, A., Bellinger, D. C., & Hauser, R. (2014). Developmental neurotoxicity of ortho-phthalate diesters: Review of human and experimental evidence. *NeuroToxicology*, *41*, 112–122. https://doi.org/10.1016/j.neuro.2014.01.007

Mitro, S. D., Johnson, T., & Zota, A. R. (2015). Cumulative Chemical Exposures During Pregnancy and Early Development. *Current Environmental Health Reports*, *2*(4), 367–378. https://doi.org/10.1007/s40572-015-0064-x

Moisiadis, V. G., & Matthews, S. G. (2014). Glucocorticoids and fetal programming part 1: outcomes. *Nature Reviews Endocrinology*, *10*(7), 391–402. https://doi.org/10.1038/nrendo.2014.73

Momotani, N., Ito, K., Hamada, N., Ban, Y., Nishikawa, Y., & Mimura, T. (1984). Maternal hyperthyroidism and congenital malformation in the offspring. *Clinical Endocrinology*, *20*(6), 695–700. https://doi.org/10.1111/j.1365-2265.1984.tb00119.x

Nakiwala, D., Peyre, H., Heude, B., Bernard, J. Y., Béranger, R., Slama, R., Philippat, C., Charles, M. A., De Agostini, M., Forhan, A., Ducimetière, P., Kaminski, M., Saurel-Cubizolles, M. J., Dargent-Molina, P., Fritel, X., Larroque, B., Lelong, N., Marchand, L., Nabet, C., … Botton, J. (2018). In-utero exposure to phenols and phthalates and the intelligence quotient of boys at 5 years. *Environmental Health: A Global Access Science Source*, *17*(1). https://doi.org/10.1186/s12940-018-0359-0

National Institute of Environmental Health Sciences. (2012). *Strategic Plan. Advancing Science, Improving Health: A Plan for Environmental Health Research 2012-2017*. https://www.niehs.nih.gov/health/materials/niehs_strategic_plan_20122017_frontiers_in_environmental_health_sciences_booklet_508.pdf

National Research Council. (1991). *Environmental Epidemiology, Volume 1: Public Health and Hazardous Wastes*. The National Academies Press. https://doi.org/10.17226/1802

Neale, D. M., Cootauco, A. C., & Burrow, G. (2007). Thyroid disease in pregnancy. *Clinics in Perinatology*, *34*(4), 543–557, v–vi. https://doi.org/10.1016/j.clp.2007.10.003

Needham, L. L., Calafat, A. M., & Barr, D. B. (2007). Uses and issues of biomonitoring. In *International Journal of Hygiene and Environmental Health* (Vol. 210, Issue 3, pp. 229–238). https://doi.org/https://doi.org/10.1016/j.ijheh.2006.11.002

Noorlander, C. W., Tijsseling, D., Hessel, E. V. S., de Vries, W. B., Derks, J. B., Visser, G. H. A., & de Graan, P. N. E. (2014). Antenatal Glucocorticoid Treatment Affects Hippocampal Development in Mice. *PLOS ONE*, *9*(1), e85671. https://doi.org/10.1371/journal.pone.0085671

Ntantu Nkinsa, P., Muckle, G., Ayotte, P., Lanphear, B. P., Arbuckle, T. E., Fraser, W. D., & Bouchard, M. F. (2020). Organophosphate pesticides exposure during fetal development and IQ scores in 3 and 4-year old Canadian children. *Environmental Research*, *190*, 110023. https://doi.org/https://doi.org/10.1016/j.envres.2020.110023

O'Connell, S. G., Kincl, L. D., & Anderson, K. A. (2014). Silicone Wristbands as Personal Passive Samplers. *Environmental Science & Technology*, *48*(6), 3327–3335. https://doi.org/10.1021/es405022f

Olesen, T. S., Bleses, D., Andersen, H. R., Grandjean, P., Frederiksen, H., Trecca, F., Bilenberg, N., Kyhl, H. B., Dalsager, L., Jensen, I. K., Andersson, A. M., & Jensen, T. K. (2018). Prenatal phthalate exposure and language development in toddlers from the Odense Child Cohort. *Neurotoxicology and Teratology*. https://doi.org/10.1016/j.ntt.2017.11.004

Oliveira, K. J., Chiamolera, M. I., Giannocco, G., Pazos-Moura, C. C., & Ortiga-Carvalho, T. M. (2019). Thyroid function disruptors: From nature to chemicals. *Journal of Molecular Endocrinology*, *62*(1), R1–R19. https://doi.org/10.1530/JME-18-0081

Oulhote, Y., Lanphear, B., Braun, J. M., Webster, G. M., Arbuckle, T. E., Etzel, T., Forget-dubois, N., Seguin, J. R., Bouchard, M. F., Macfarlane, A., Ouellet, E., Fraser, W., & Muckle, G. (2020). *Gestational Exposures to Phthalates and Folic Acid, and Autistic Traits in Canadian*. *128*(February), 1–12.

Pacyga, D. C., Sathyanarayana, S., & Strakovsky, R. S. (2019). Dietary Predictors of Phthalate and Bisphenol Exposures in Pregnant Women. *Advances in Nutrition*. https://doi.org/10.1093/advances/nmz029

Page, B. D., & Lacroix, G. M. (1995). The occurrence of phthalate ester and di-2-ethylhexyl adipate plasticizers in canadian packaging and food sampled in 1985-1989: A survey. *Food Additives and Contaminants, 12*(1), 129–151. https://doi.org/10.1080/02652039509374287

Papadopoulou, E., Haug, L. S., Sakhi, A. K., Andrusaityte, S., Basagaña, X., Brantsaeter, A. L., Casas, M., Fernández-Barrés, S., Grazuleviciene, R., Knutsen, H. K., Maitre, L., Meltzer, H. M., McEachan, R. R. C., Roumeliotaki, T., Slama, R., Vafeiadi, M., Wright, J., Vrijheid, M., Thomsen, C., & Chatzi, L. (2019). Diet as a source of exposure to environmental contaminants for pregnant women and children from Six European Countries. *Environmental Health Perspectives, 127*(10), 1–13. https://doi.org/10.1289/EHP5324

Parlett, L. E., Calafat, A. M., & Swan, S. H. (2013). Women's exposure to phthalates in relation to use of personal care products. *Journal of Exposure Science and Environmental Epidemiology, 23*(2), 197–206. https://doi.org/10.1038/jes.2012.105

Patnaik, P. (2017). *Handbook of environmental analysis : chemical pollutants in air, water, soil, and solid wastes* (Third edit). Boca Raton : Taylor & Francis, CRC Press, 2017.

Peck, J. D., Sweeney, A. M., Symanski, E., Gardiner, J., Silva, M. J., Calafat, A. M., & Schantz, S. L. (2010). Intra- and inter-individual variability of urinary phthalate metabolite concentrations in Hmong women of reproductive age. *Journal of Exposure Science & Environmental Epidemiology, 20*(1), 90–100. https://doi.org/10.1038/jes.2009.4

Pekkanen, J., & Pearce, N. (2001). Environmental epidemiology: challenges and opportunities. *Environmental Health Perspectives, 109*(1), 1–5. https://doi.org/10.1289/ehp.011091

Perrier, F., Giorgis-Allemand, L., Slama, R., & Philippat, C. (2016). Within-subject Pooling of Biological Samples to Reduce Exposure Misclassification in Biomarker-based Studies. *Epidemiology (Cambridge, Mass.), 27*(3), 378–388. https://doi.org/10.1097/EDE.0000000000000460

Philippat, C., Bennett, D., Calafat, A. M., & Picciotto, I. H. (2015). Exposure to select phthalates and phenols through use of personal care products among Californian adults and their children. *Environ Res, 140*, 369–376.

Polanska, K., Ligocka, D., Sobala, W., & Hanke, W. (2014). Phthalate exposure and child development: The Polish Mother and Child Cohort Study. *Early Human Development, 90*(9), 477–485. https://doi.org/https://doi.org/10.1016/j.earlhumdev.2014.06.006

Poon, R., Lecavalier, P., Mueller, R., Valli, V. E., Procter, B. G., & Chu, I. (1997). Subchronic oral toxicity of di-n-octyl phthalate and di(2-ethylhexyl) phthalate in the rat. *Food and Chemical Toxicology, 35*(2), 225–239. https://doi.org/10.1016/S0278-6915(96)00064-6

Qian, X., Li, J., Xu, S., Wan, Y., Li, Y., Jiang, Y., Zhao, H., Zhou, Y., Liao, J., Liu, H., Sun, X., Liu, W., Peng, Y., Hu, C., Zhang, B., Lu, S., Cai, Z., & Xia, W. (2019). Prenatal exposure to phthalates and neurocognitive development in children at two years of age. *Environment International*. https://doi.org/10.1016/j.envint.2019.105023

Radke, E. G., Braun, J. M., Nachman, R. M., Cooper, G. S., Agency, U. S. E. P., & States, U. (2020). Phthalate exposure and neurodevelopment : A systematic review and meta-analysis of human epidemiological evidence. *Environment International, 137*(September 2019), 105408. https://doi.org/10.1016/j.envint.2019.105408

Reid, A. M., Brougham, C. A., Fogarty, A. M., & Roche, J. J. (2007). An investigation into possible sources of phthalate contamination in the environmental analytical laboratory. *International Journal of Environmental Analytical Chemistry, 87*(2), 125–133. https://doi.org/10.1080/03067310601071183

Reiss, A. L., Abrams, M. T., Singer, H. S., Ross, J. L., & Denckla, M. B. (1996). Brain development, gender and IQ in children. A volumetric imaging study. *Brain : A Journal of Neurology, 119 (Pt 5,* 1763–1774. https://doi.org/10.1093/brain/119.5.1763

Ribas-Fitó, N., Sala, M., Kogevinas, M., & Sunyer, J. (2001). Polychlorinated biphenyls (PCBs) and neurological development in children: a systematic review. *Journal of Epidemiology and Community Health, 55*(8), 537–546. https://doi.org/10.1136/jech.55.8.537

Rodríguez-Carmona, Y., Ashrap, P., Calafat, A. M., Ye, X., Rosario, Z., Bedrosian, L. D., Huerta-Montanez, G., Vélez-Vega, C. M., Alshawabkeh, A., Cordero, J. F., Meeker, J. D., & Watkins, D. (2020). Determinants and characterization of exposure to phthalates, DEHTP and DINCH among pregnant women in the PROTECT birth cohort in Puerto Rico. *Journal of Exposure Science & Environmental Epidemiology, 30*(1), 56–69. https://doi.org/10.1038/s41370-019-0168-8

Romano, M. E., Eliot, M. N., Zoeller, R. T., Hoofnagle, A. N., Calafat, A. M., Karagas, M. R., Yolton, K., Chen, A., Lanphear, B. P., & Braun, J. M. (2018). Maternal urinary phthalate metabolites during pregnancy and thyroid hormone concentrations in maternal and cord sera: The HOME Study. In *International Journal of Hygiene and Environmental Health* (Vol. 221, Issue 4, pp. 623–631). Elsevier. https://doi.org/10.1016/j.ijheh.2018.03.010

Rothman, K. J. (1993). Methodologic frontiers in environmental epidemiology. *Environmental Health Perspectives, 101 Suppl*(Suppl 4), 19–21. https://doi.org/10.1289/ehp.93101s419

Samandar, E., Silva, M. J., Reidy, J. A., Needham, L. L., & Calafat, A. M. (2009). Temporal stability of eight phthalate metabolites and their glucuronide conjugates in human urine. *Environmental Research, 109*(5), 641–646. https://doi.org/https://doi.org/10.1016/j.envres.2009.02.004

Saravanabhavan, G., Guay, M., Langlois, É., Giroux, S., Murray, J., & Haines, D. (2013). *Canadian Health Measures Survey (2007 – 2009).* 1–10.

Schecter, A., Lorber, M., Guo, Y., Wu, Q., Yun, S. H., Kannan, K., Hommel, M., Imran, N., Hynan, L. S., Cheng, D., Colacino, J. A., & Birnbaum, L. S. (2013). Phthalate concentrations and dietary exposure from food purchased in New York State. *Environmental Health Perspectives, 121*(4), 473–494. https://doi.org/10.1289/ehp.1206367

Schettler, T., Skakkebæk, N., De Kretser, D., & Leffers, H. (2006). Human exposure to phthalates via consumer products. *International Journal of Andrology, 29*(1), 134–139. https://doi.org/10.1111/j.1365-2605.2005.00567.x

Schisterman, E. F., Cole, S. R., & Platt, R. W. (2009). Overadjustment bias and unnecessary adjustment in epidemiologic studies. *Epidemiology (Cambridge, Mass.), 20*(4), 488–495. https://doi.org/10.1097/EDE.0b013e3181a819a1

Schlumpf, M., Kypke, K., Wittassek, M., Angerer, J., Mascher, H., Mascher, D., Vokt, C., Birchler, M., & Lichtensteiger, W. (2010). Exposure patterns of UV filters, fragrances, parabens, phthalates, organochlor pesticides, PBDEs, and PCBs in human milk: correlation of UV filters with use of cosmetics. *Chemosphere, 81*(10), 1171–1183. https://doi.org/10.1016/j.chemosphere.2010.09.079

Serrano, S. E., Braun, J., Trasande, L., Dills, R., & Sathyanarayana, S. (2014). Phthalates and diet: A review of the food monitoring and epidemiology data. *Environmental Health: A Global Access Science Source, 13*(1). https://doi.org/10.1186/1476-069X-13-43

Shaffer, C. B., Carpenter, C. P., & Smyth Jr., H. F. (1945). Acute and Subacute Toxicity of Di (2-Ethylhexyl) Phthalate with Note upon its Metabolism. *Journal of Industrial Hygiene and Toxicology, 27*(5), 130–135.

Shepard, T. H. (1967). Onset of function in the human fetal thyroid: biochemical and radioautographic studies from organ culture. *J Clin Endocrinol Metab, 27*(7), 945–958. https://doi.org/10.1210/jcem-27-7-945

Shu, H., Jönsson, B. A. G., Gennings, C., Lindh, C. H., Nånberg, E., & Bornehag, C.-G. (2019). PVC flooring at home and uptake of phthalates in pregnant women. *Indoor Air, 29*(1), 43–54. https://doi.org/10.1111/ina.12508

Shu, H., Jonsson, B. A., Gennings, C., Svensson, A., Nanberg, E., Lindh, C. H., Knutz, M., Takaro, T. K., & Bornehag, C.-G. (2018). Temporal trends of phthalate exposures during 2007-2010 in Swedish pregnant women. *Journal of Exposure Science & Environmental Epidemiology, 28*(5), 437–447. https://doi.org/10.1038/s41370-018-0020-6

Shuttleworth-Edwards, A. B. (2016). Generally representative is representative of none: commentary on the pitfalls of IQ test standardization in multicultural settings. *The Clinical Neuropsychologist, 30*(7), 975–998. https://doi.org/10.1080/13854046.2016.1204011

Silva, M., Wong, L.-Y., Samandar, E., Preau, J., Jia, L., & Calafat, A. (2019). Exposure to di-2-ethylhexyl terephthalate in the U.S. general population from the 2015–2016 National Health and Nutrition Examination Survey. *Environment International, 123*, 141–147. https://doi.org/https://doi.org/10.1016/j.envint.2018.11.041

Silva, M J, Reidy, J. A., Herbert, A. R., Preau, J. L., Needham, L. L., & Calafat, A. M. (2004). Detection of Phthalate Metabolites in Human Amniotic Fluid. *Bulletin of Environmental Contamination and Toxicology, 72*(6), 1226–1231. https://doi.org/10.1007/s00128-004-0374-4

Silva, Manori J, Jia, T., Samandar, E., Preau, J. L. J., & Calafat, A. M. (2013). Environmental exposure to the plasticizer 1,2-cyclohexane dicarboxylic acid, diisononyl ester (DINCH) in U.S. adults (2000-2012). *Environmental Research, 126*, 159–163. https://doi.org/10.1016/j.envres.2013.05.007

Sörberg Wallin, A., Allebeck, P., Gustafsson, J.-E., & Hemmingsson, T. (2018). Childhood IQ and mortality during 53 years' follow-up of Swedish men and women. *Journal of Epidemiology and Community Health, 72*(10), 926–932. https://doi.org/10.1136/jech-2018-210675

Stafoggia, M, Breitner, S., Hampel, R., & Basagana, X. (2017). *Statistical Approaches to Address Multi-Pollutant Mixtures and Multiple Exposures: the State of the Science*.

Stafoggia, Massimo, Breitner, S., Hampel, R., & Basagaña, X. (2017). Statistical Approaches to Address

Multi-Pollutant Mixtures and Multiple Exposures: the State of the Science. *Current Environmental Health Reports*, *4*(4), 481–490. https://doi.org/10.1007/s40572-017-0162-z

Staples, C. A., Peterson, D. R., Parkerton, T. F., & Adams, W. J. (1997). The environmental fate of phthalate esters: A literature review. *Chemosphere*, *35*(4), 667–749. https://doi.org/10.1016/S0045-6535(97)00195-1

Statistics Canada. (2020). *Canadian Health Measures Survey (CHMS) January 2018 to December 2019 (Cycle 6)*. https://www23.statcan.gc.ca/imdb/p2SV.pl?Function=getSurvey&SDDS=5071&lang=en&db=imdb &adm=8&dis=2

Steinley, D. (2006). K-means clustering: a half-century synthesis. *The British Journal of Mathematical and Statistical Psychology*, *59*(Pt 1), 1–34. https://doi.org/10.1348/000711005X48266

Sugiyama, S., Shimada, N., Miyoshi, H., & Yamauchi, K. (2005). Detection of Thyroid Systemâ€"Disrupting Chemicals Using in Vitro and in Vivo Screening Assays in Xenopus laevis. *Toxicological Sciences*, *88*(2), 367–374. https://doi.org/10.1093/toxsci/kfi330

Sun, D., Zhou, L., Wang, S., Liu, T., Zhu, J., Jia, Y., Xu, J., Chen, H., Wang, Q., & Xu, F. (2018). Effect of Di-(2-ethylhexyl) phthalate on the hypothalamus-pituitary-thyroid axis in adolescent rat. *Endocrine Journal*, *65*, 261–268. https://doi.org/10.1507/endocrj.EJ17-0272

Swan, S. H., Liu, F., Hines, M., Kruse, R. L., Wang, C., Redmon, J. B., Sparks, A., & Weiss, B. (2010). Prenatal phthalate exposure and reduced masculine play in boys. *Int J Androl*, *33*(2), 259–269.

Tanner, E. M., Bornehag, C.-G., & Gennings, C. (2019). Repeated holdout validation for weighted quantile sum regression. *MethodsX*, *6*, 2855–2860. https://doi.org/https://doi.org/10.1016/j.mex.2019.11.008

Tanner, E, Hallerbäck, M., Wikström, S., Lindh, C., Kiviranta, H., Gennings, C., & Bornehag, C.-G. (2019). Early prenatal exposure to suspected endocrine disruptor mixtures is associated with lower IQ at age seven. In *Environment International* (p. 105185). https://doi.org/https://doi.org/10.1016/j.envint.2019.105185

Tanner, Eva, Lee, A., & Colicino, E. (2020). Environmental mixtures and children's health: identifying appropriate statistical approaches. *Current Opinion in Pediatrics*, *32*(2), 315–320. https://doi.org/10.1097/MOP.0000000000000877

Taylor, K. W., Joubert, B. R., Braun, J. M., Dilworth, C., Gennings, C., Hauser, R., Heindel, J. J., Rider, C. V, Webster, T. F., & Carlin, D. J. (2016). Statistical Approaches for Assessing Health Effects of Environmental Chemical Mixtures in Epidemiology: Lessons from an Innovative Workshop. *Environ Health Perspect*, *124*(12), A227–A229.

Téllez-Rojo, M. M., Cantoral, A., Cantonwine, D. E., Schnaas, L., Peterson, K., Hu, H., & Meeker, J. D. (2013). Prenatal urinary phthalate metabolites levels and neurodevelopment in children at two and three years of age. *Sci Total Environ*, *0*, 386–390.

Testa, C., Nuti, F., Hayek, J., De Felice, C., Chelli, M., Rovero, P., Latini, G., & Papini, A. M. (2012). Di-(2-ethylhexyl) phthalate and autism spectrum disorders. *ASN Neuro*, *4*(4), 223–229. https://doi.org/10.1042/AN20120015

Tibshirani, R. (1996). Regression Shrinkage and Selection via the Lasso. *Journal of the Royal Statistical Society. Series B (Methodological)*, *58*(1), 267–288. http://www.jstor.org/stable/2346178

Toivonen, K. I., Oinonen, K. A., & Duchene, K. M. (2017). Preconception health behaviours: A scoping review. *Preventive Medicine*, *96*, 1–15. https://doi.org/10.1016/j.ypmed.2016.11.022

Tranfo, G., Caporossi, L., Pigini, D., Capanna, S., Papaleo, B., & Paci, E. (2018). Temporal Trends of Urinary Phthalate Concentrations in Two Populations: Effects of REACH Authorization after Five Years. *International Journal of Environmental Research and Public Health*, *15*(9), 1950. https://doi.org/10.3390/ijerph15091950

Trasande, L., Zoeller, R. T., Hass, U., Kortenkamp, A., Grandjean, P., Myers, J. P., DiGangi, J., Bellanger, M., Hauser, R., Legler, J., Skakkebaek, N. E., & Heindel, J. J. (2015). Estimating Burden and Disease Costs of Exposure to Endocrine-Disrupting Chemicals in the European Union. *The Journal of Clinical Endocrinology & Metabolism*, *100*(4), 1245–1255. https://doi.org/10.1210/jc.2014-4324

U.S. Consumer Product Safety Commission. (2014). *Chronic Hazard Advisory Panel on Phthalates and Phthalate Alternatives*. https://www.cpsc.gov/en/Regulations-Laws--Standards/Statutes/The-Consumer-Product-Safety-Improvement-Act/Phthalates/Chronic-Hazard-Advisory-Panel-CHAP-on-Phthalates/

van den Dries, M. A., Pronk, A., Guxens, M., Spaan, S., Voortman, T., Jaddoe, V. W., Jusko, T. A., Longnecker, M. P., & Tiemeier, H. (2018). Determinants of organophosphate pesticide exposure in pregnant women: A population-based cohort study in the Netherlands. *International Journal of Hygiene and Environmental Health*. https://doi.org/10.1016/j.ijheh.2018.01.013

Verner, M.-A., Salame, H., Housand, C., Birnbaum, L. S., Bouchard, M. F., Chevrier, J., Aylward, L. L., Naiman, D. Q., & LaKind, J. S. (2020). How Many Urine Samples Are Needed to Accurately Assess Exposure to Non-Persistent Chemicals? The Biomarker Reliability Assessment Tool (BRAT) for Scientists, Research Sponsors, and Risk Managers. *International Journal of Environmental Research and Public Health*, *17*(23). https://doi.org/10.3390/ijerph17239102

Vohr, B. R., Poggi Davis, E., Wanke, C. A., & Krebs, N. F. (2017). Neurodevelopment: The Impact of Nutrition and Inflammation During Preconception and Pregnancy in Low-Resource Settings. *Pediatrics*, *139*(Suppl 1), S38–S49. https://doi.org/10.1542/peds.2016-2828F

Warrilow, A., Der, G., Cooper, S.-A., Minnis, H., & Pell, J. P. (2021). Childhood neurodevelopmental markers and risk of premature mortality: Follow-up to age 60-65 years in the Aberdeen Children of the 1950s study. *PloS One*, *16*(8), e0255649. https://doi.org/10.1371/journal.pone.0255649

Weiss, B. (2012). The intersection of neurotoxicology and endocrine disruption. *Neurotoxicology*, *33*(6), 1410–1419. https://doi.org/10.1016/j.neuro.2012.05.014

Weng, H.-Y., Hsueh, Y.-H., Messam, L. L. M., & Hertz-Picciotto, I. (2009). Methods of Covariate Selection: Directed Acyclic Graphs and the Change-in-Estimate Procedure. *American Journal of Epidemiology*, *169*(10), 1182–1190. https://doi.org/10.1093/aje/kwp035

Whyatt, R. M., Liu, X., Rauh, V. A., Calafat, A. M., Just, A. C., Hoepner, L., Diaz, D., Quinn, J., Adibi, J., Perera, F. P., & Factor-Litvak, P. (2012). Maternal prenatal urinary phthalate metabolite concentrations and child mental, psychomotor, and behavioral development at 3 years of age. *Environmental Health Perspectives*, *120*(2), 290–295. https://doi.org/10.1289/ehp.1103705

Wild, C. P. (2005). Complementing the Genome with an "Exposome": The Outstanding Challenge of Environmental Exposure Measurement in Molecular Epidemiology. *Cancer Epidemiology Biomarkers & Prevention, 14*(8), 1847. http://cebp.aacrjournals.org/content/14/8/1847.abstract

Wild, C. P. (2012). The exposome: from concept to utility. *International Journal of Epidemiology, 41*(1), 24–32. https://doi.org/10.1093/ije/dyr236

Wilkinson, P. O. (2006). *Environmental epidemiology* (P. Wilkinson (ed.)). Open University Press.

Wood, C. E. (2013). Development and programming of the hypothalamus-pituitary-adrenal axis. *Clinical Obstetrics and Gynecology, 56*(3), 610–621. https://doi.org/10.1097/GRF.0b013e31829e5b15

Woods, M. M., Lanphear, B. P., Braun, J. M., & McCandless, L. C. (2017). Gestational exposure to endocrine disrupting chemicals in relation to infant birth weight: a Bayesian analysis of the HOME Study. *Environmental Health, 16*(1), 115. https://doi.org/10.1186/s12940-017-0332-3

Wraw, C., Deary, I. J., Gale, C. R., & Der, G. (2015). Intelligence in youth and health at age 50. *Intelligence, 53*, 23–32. https://doi.org/10.1016/j.intell.2015.08.001

Xiang, W., Gong, Q., Xu, J., Li, K., Yu, F., Chen, T., Qin, S., Li, C., & Wang, F. (2020). Cumulative risk assessment of phthalates in edible vegetable oil consumed by Chinese residents. *Journal of the Science of Food and Agriculture, 100*(3), 1124–1131. https://doi.org/10.1002/jsfa.10121

Yao, H., Han, Y., Gao, H., Huang, K., Ge, X., Xu, Y., Xu, Y., Jin, Z., Sheng, J., Yan, S., Zhu, P., Hao, J., & Tao, F. (2016). Maternal phthalate exposure during the first trimester and serum thyroid hormones in pregnant women and their newborns. *Chemosphere, 157*, 42–48. https://doi.org/https://doi.org/10.1016/j.chemosphere.2016.05.023

Zhang, Q., Sun, Y., Zhang, Q., Hou, J., Wang, P., Kong, X., & Sundell, J. (2020). Phthalate exposure in Chinese homes and its association with household consumer products. *Science of The Total Environment, 719*, 136965. https://doi.org/https://doi.org/10.1016/j.scitotenv.2020.136965

Zhang, Y., Dong, T., Hu, W., Wang, X., Xu, B., Lin, Z., Hofer, T., Stefanoff, P., Chen, Y., Wang, X., & Xia, Y. (2019). Association between exposure to a mixture of phenols, pesticides, and phthalates and obesity: Comparison of three statistical models. In *Environment International* (Vol. 123, pp. 325–336). https://doi.org/https://doi.org/10.1016/j.envint.2018.11.076

Zhao, H., Li, J., Zhou, Y., Zhu, L., Zheng, Y., Xia, W., Li, Y., Xiang, L., Chen, W., Xu, S., & Cai, Z. (2018). Investigation on Metabolism of Di(2-Ethylhexyl) Phthalate in Different Trimesters of Pregnant Women. *Environmental Science & Technology, 52*(21), 12851–12858. https://doi.org/10.1021/acs.est.8b04519

Zota, A., Calafat, A., & Woodruff, T. (2014). Temporal trends in phthalate exposures: findings from the National Health and Nutrition Examination Survey, 2001-2010. *Environmental Health Perspectives, 122*(3), 235–241.

Zota, A. R., Phillips, C. A., & Mitro, S. D. (2016). Recent Fast Food Consumption and Bisphenol A and Phthalates Exposures among the U.S. Population in NHANES, 2003–2010. *Environmental Health Perspectives, 124*(10), 1521–1528. https://doi.org/10.1289/ehp.1510803

Zou, H., & Hastie, T. (2005). Regularization and Variable Selection via the Elastic Net. *Journal of the Royal Statistical Society. Series B (Statistical Methodology)*, *67*(2), 301–320. http://www.jstor.org/stable/3647580